Messages from Earth

Messages from Earth

Nature and the Human Prospect in Alaska

Robert B. Weeden

University of Alaska Press
Fairbanks, 1992

Library of Congress Cataloging-in-Publication Data

Weeden, Robert B., 1933-
        Messages from earth : nature and the human prospect in Alaska /
    Robert B. Weeden.
        p. cm.
        Includes bibliographical references and index.
        ISBN 0-912006-56-0
        1. Ecology--Alaska. 2. Human ecology--Alaska. 3. Natural
    resources--Alaska. I. Title
    QH105.A4W44 1992   91-43178
    333.7'09798--dc20   CIP

International Standard Book Number: 0-912006-56-0
Library of Congress Catalogue Number: 91-43178

Printed in the United States by: Thomson-Shore, Inc.
        on 60# Glatfelter Thor Offset paper, recycled and acid-free.

This publication was printed on acid-free paper which meets the minimum
requirements for American National Standard for Information Sciences—
Permanence of Paper for Printed Library Materials, ANSI Z39.48-1984.

Book and cover design by Dixon Jones, maps by Eve Griffin, IMPACT
        Graphics, University of Alaska Fairbanks.
Cover photograph by Nancy Simmerman/AllStock, Inc.
Fern illustration by Russell Mitchell
Production and publication coordination by Deborah Van Stone.

To Larry Bright, Nora Kelly and Kent Patrick-Riley,
who walked much of this way with me.

Contents

Preface

Yogi Berra once said, "You can observe a lot just by watching." I have watched Alaska for three decades, its countryside and humanscape moving together in an infinitely intricate dance. Seasons have come and gone only to return almost—but not ever quite—to their remembered place. Woods have grown up and burned down, salmon runs waxed and waned, glaciers surged and fallen back. Roads have been freshly built, crumbled and patched. Houses have opened virginal doors, absorbed laughter and tears, shifted on thawing ground, heard the final closing of their doors. Good ideas have turned into successful businesses, bad ideas into debts, grand schemes into embarrassments, earnest plans into muddles.

Recently I have looked for patterns in the dance. Do people seem alert to the landscape they live in? Are we working with nature's grain, or against it? Is there a sense of belonging as well as possessing? Do we behave as if tomorrow matters? Are we building toward a sustainable society?

This book grew out of that search. Its focus is Alaska. The problem, though, is global. To my knowledge no part of the modern world—tropical, temperate or arctic; industrial or pastoral; capitalist or communist—has built an enduring, person-enhancing society sustained by its home landscape. Perhaps no one part of the world can achieve such a society very far ahead of the others, given the enormous world traffic in people, ideas, debts and goods. Because of the universality of the issues, parts of this book may be useful to others who, in their place of living, work toward that integration. No, it is not because Alaska is unique that I think and write about it. It *is* unique, but so is every other place. And every place needs to be absorbed, distilled, interpreted, its story told and retold until it is everyone's.

As I think of the enormity of Alaska as geographic space, as a multi-millennial history of human striving, and as a small but complex and changeful contemporary society, I wonder about my own foolhardiness for trying to think about even so limited a piece of this humanized Earth. My time in Alaska is little more than one-thousandth of the span of human living in the North, and I know only a little of what has happened even in that fraction of Alaska's history. The three deceptively simple questions Wendell Berry[1] asked in "Home Economics"—What is here? What will nature permit us to do here? What will nature help us to do here?—cannot be answered truly and practically for a "here" the size and complexity of Alaska. The real work must come from those who bound themselves much more modestly.

There is something about Alaska that makes the impossible seem worth trying. The North is mythic, a place of heroic adventure, the place we wanted to go to when we were younger. It has a feeling of freshness and open possibility. For many who live and work in Alaska there is a powerful sense that what one does, matters. In this small place a person can leave a mark.

If we hold that self-image of powerfulness, if we believe we can help shape futures, we also must pause to ask whether we know where and how to use that capacity. In a society more characterized by action than contemplation, the temptation to respond only to the apparent demands of the present is always strong. This is true especially of restive, transient societies, of which Alaska's is one of many.

It seems to me that to do things that are consistent with the strengths, capacities and limitations of local environments is a reasonable, practical and important guide to action, however bewildering the social context may be at the moment. It is a guide that can serve whether we are crafting grand designs or just muddling through. To know the home country is crucial. It should become second nature.

Acknowledgments

This book was a much-interrupted project beginning in 1981 and ending, a half dozen drafts later, in 1991. To the many typists who converted my poor penmanship into readable typescript, and to the department heads who seemed miraculously understanding when, to their annual question, I always replied, "Still working on it," I extend my gratitude. My particular thanks go to Ms. Tina Picolo who, while producing the final two versions, had the sense to consult manuals of style rather than to rely on my ingrained writing prejudices.

Any merit in this book arises from the trial-and-error process of testing ideas, dipped from a broad river of literature on environment and resource management, against the perceived realities of Alaskan landscapes, people and issues. Crucial help in that process came from graduate and undergraduate students at the University of Alaska Fairbanks. I would venture a word-picture of how a situation had arisen or could be resolved, and the ensuing discussion never failed to clarify my comprehension and improve (occasionally by early burial) my "solution." At least a dozen books I came to cherish and draw from were initially pointed out to me by these student colleagues. Though I have dedicated this book to three who enlivened my mind during the pivotal years 1984-86, a score of others contributed as much in ways whose details I've now forgotten. To all of them, my deepest respect and boundless thanks.

I might well have given up this project before the end of its lurching ten-year journey, except for Judy Weeden's steadfast faith that it was worth doing. She may have second thoughts when she sees the final product, but that will be a failing of my execution, not of her loving encouragement.

Introduction

It is well to be informed about the winds,

About the variations in the sky,

The native traits and habits of the place,

What each locale permits, and what denies.

—Virgil, *The Georgics*

Biologically, we are still a subtropical woodland species. Our global peregrinations during the past hundred centuries have affected our genetic inheritance only very modestly. Humankind's remarkable colonization of nearly all of Earth's starkly contrasting landscapes has been accomplished by modifying tools and behavior to meet the requirements of unfamiliar environments, and by making pristine environments more benign (from our human viewpoint) and productive.

The two obvious advantages of cultural over genetic evolution are the former's speed and great flexibility. Equipped with simple tools and skills, *Homo sapiens* spread during a few thousand years into temperate, island and polar regions. During a more recent time of written history, the discovery, conquest and settlement of North America, Australia and Oceania by Europeans took only a few hundred years. During my lifetime a permanent settlement was established on Antarctica almost as soon as the decision was made to do so. Already the now-familiar fight between preserving nature and exploiting resources has begun there. Springing from earlier achievements of atmospheric flight and electronics, humans learned in less than two decades to fly to the moon and back again. Mars is next. With our mechanical surrogates aboard, Voyager II has reported on close views of Neptune; Galileo now probes toward Jupiter.

However rapidly learned adaptations have allowed humans to invade diverse environments, cultural evolution has yet to prove its ability to produce long-term staying power. Time after time particular social patterns and technologies have abruptly disappeared. Some took too much for too long from Nature or required extensive manipulations of landscapes that finally led to the collapse and restructuring of natural systems. Some were overrun by an invading culture, which might have been either less or more "advanced" than the defeated ones. Still others were convulsed from within by technical or social revolution.

Probably the most universal failing of recent cultural evolution is its inability to adjust its numerous forms to the rhythms, character and physical limitations of regional environments. The human

3

forte has been the development of exploitive capacity. Perhaps the very thing of which humanity is so proud—the ability to change social and technological systems rapidly—is inherently antiadaptive, imposed as it is on a natural world with vastly slower evolutionary or adaptive response times. Today's warnings of global climatic changes caused unwittingly by people are an example. The frightful cost, in human habitat and lives, of the modernization of Amazonia is another case in point. Where is the society today whose resource demands are no more than sustainable local supply? Where is the culture that is building its soil, cherishing and replenishing its waters, purifying its air? Where is the modern nation with wealthy cities and pleasant countrysides that is not parasitizing distant landscapes?

I think—certainly hope—that humanity is beginning to realize that exploitive societies have had their day. They took us from beginnings in Africa to every corner of every continent, from caves and thickets to cities of glass and steel. They once encompassed populations of tens of thousands, and now count their billions. Although some enthusiasts assert that we can support even more billions in even more places by digging deeper and flying higher, many people wonder what is being proven, and at what risk of disaster? The true frontier, it seems to many of us, is better, not more; congruence, not combat, with nature and with each other. This does not imply a denial of knowledge or a return to the primitive. It means converting knowledge into wisdom, destructive technology into benign tools, transient exploitation into durable residence.

People of many nations, for example those from more than 100 nations who, in 1978 and 1979, prepared the World Conservation Strategy[2] and who in the years since have carried it forward, are convinced that a dramatic reorientation of thought and behavior is crucial to human survival. Yet, few nations are ready to act on that conviction. Poverty and explosively rising populations force many tropical nations into ever-increasing destruction of forests and exploitation of water, soil and mineral resources. Other nations, mainly in temperate zones, still are wealthy enough to buy, or powerful enough to control, a continued stream of resource imports. Not a few nations, scrambling to maintain high consumption and dominated by interests that fear a significant societal reorientation, simply will not admit the problem. This is a time of vacillation and violent uncertainties, a behavioral watershed.

Alaska, although one of 50 United States, stands in singular relationship to this watershed. It has the smallest population, the largest area, the most pristine landscapes, and among the greatest per capita public and private income of any state. Its freedom of choice seems extraordinary.

But on whom does the responsibility for these choices fall? About three-fourths of all Alaskans are newly arrived from the south ("Outside"), either during the last year, the last decade, or the last generation. Many do not plan to stay. They will be replaced by other newcomers. The worldviews and behavior of these quasi-residents are rooted in industrialized, urbanized, mid-century America. Their expectations of nature have little relationship to the ice-dominated seas, glacier-molded rivers and permafrost-locked soils typical of all but a small part of Alaska. They are strangers in a strange land.

The 80,000 or so native Alaskans, descendants of the very first North Americans, recently have adopted the hardware and institutional structures of industrialized society. They can no longer use ancient northern wisdom to full advantage. In school and on the streets they are educated into the same pattern of ignorance and error as the youngsters of contemporary urban America.

All Alaskans must come to terms with the central problem of how a contemporary society can function harmoniously within arctic and subarctic environments. Harmonious function implies a deep and true knowledge of those environments and of human interactions with them. We do know something about the North. There are, however, several kinds of knowledge. On one hand are the achievements of modern science. This kind of knowledge is gained by a very small group isolated from the vast majority of society by its special skills and special style of living. The bridge between knowledge and action is a precarious one of printed words and sound bites on television. Those who must act are not scientists, and they must somehow be convinced that what the specialists have learned is useful. Most of what scientists learn is meaningless to the vast majority. At best—or worst—it feeds an equally specialized and arcane technology industry.

A totally different kind of knowledge dominated Alaska life before Europeans and modern science arrived. Native people knew their environments by direct, repeated, personal observation, encouraged by social values placed on that knowledge and sharpened by the needs of daily survival. The accumulation and transference of knowledge throughout the community came through art, dance, myth, story, the talk of the elders, the precise design of tools and crafts and small structures. Such knowledge, because it is not objectified, is intimately held and automatically incorporated into belief and behavior. And, because it is not held at arm's length or mechanistically interpreted, it cannot lead easily to complex technological development.

The ancient knowledge suited subsistence-based societies wonderfully well. Today it seems anachronistic. Few have time to listen to the old gentle myths, or to experience personally the hour-to-hour changes in a river when the ice is breaking up. It has been hard to adapt the old intimate knowledge to the forms and demands of environmental impact statements and joint venture agreements with oil and mining companies. On the other hand, modern science has not performed very well even in its own mother culture and temperate home landscape. It is too cold, too complex, too alienated, too easily captured for private profit or political power. A new knowledge system seems needed in the North. Perhaps the time is right for an intimate science to emerge here, where the slate is only newly smudged and the opportunities are so grand.

However, not even the most isolationistic Alaskans expect to be left alone in quiet contemplation of our own affairs and future. Korea and Taiwan want our coal and potential grain output. Japan, Korea, the U.S.S.R. and Poland, among others, crave (and mostly catch) Alaska's fish. Japan controls our timber industry and eyes our copper, iron and other mineral deposits. Over 98 percent of the ocean area within 200 miles of Alaska, and 60 percent of the state's land area, are in the hands of Congress and presidents whose decisions must be acceptable to a majority of some 230 million Americans, who covet (among other things) Alaskan petroleum and wilderness. Among Alaskans are many whose personal and corporate interests lie with the expansion of the resource demands of these distant people: witness the fight by Alaska's political leaders to allow foreign export of Alaskan oil, the assiduous visits of Alaskan business people to Asia and the Soviet Union to drum up trade. There wouldn't be as much business if we tended to our own business.

The upshot is that deep and pervasive changes in human understanding and behavior are required if we are to create a durable northern society. Such changes will involve communities, institutions of governance, science, the economy, and most important, the orientation of every person toward nature. The changes will be slow and no gains secure. We will not be left to strive undisturbed. The larger world will infiltrate, seduce, suggest and impose; it will change major strands of the pattern to be woven. Success will be measured by resilience in the face of external as well as internal stress. There will be no arrival at a goal, only continuing movement.

This book is a notation of part of that journey. In the first chapters I survey the seascapes and landscapes in which Alaskan endeavors are dynamically imbedded and try to find both specific and general features that could guide our design toward sustainability. Then I ask whether we seem today to be adapting well or poorly—or both at once—to arctic and subarctic life. This section includes a quick look at several activities such as mining and logging that exemplify our current relationships with nature. The last part of the book discusses strategies that may help people develop patterns of thinking and doing, and social institutions and economies, that are congruent with the character of the North. By this I mean that they must be resilient in the face of stress, opportunistic, broadly capable rather than specialized, inventively conservative of energy and materials, and wisely responsive to diversity and change. In the end I return to my own personal explorations, to try to connect nature outward and nature inward.

One way to be able to keep promises—not always the best way, I will add, but appropriate for authors of books like this one—is not to promise too much. I am not drawing a road map to Alaska's future. The future has no known locations to aim for, no neatly separable paths to mutually exclusive end points. Indeed, I do have a sense of qualities of a future that seem good, preferable to others and worth striving for. But these spring out of what seems basic and right in the present, worthy of nurturing and expanding. To live amidst diversity and choice. To have other life companionably around us. To be able to draw sustenance, energy and materials from the land and sea in fair confidence that we are not demanding more than a resilient but not indestructible Nature can give. To do so skillfully, without waste, in joy and satisfaction at doing a job well.

Part One

This Place Called Alaska

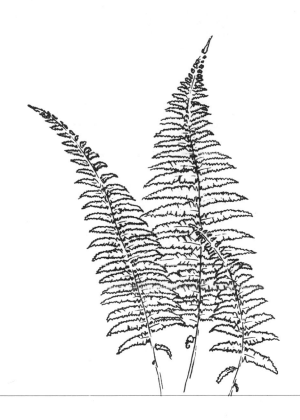

THE GIFT OUTRIGHT

The land was ours before we were the land's.
She was our land more than a hundred years
Before we were her people. She was ours
In Massachusetts, in Virginia.
But we were England's, still colonials,
Possessing what we still were unpossessed by,
Possessed by what we now no more possessed.
Something we were withholding made us weak
Until we found out that it was ourselves
We were withholding from our land of living,
And forthwith found salvation in surrender.
Such as we were we gave ourselves outright
(The deed of gift was many deeds of war)
To the land vaguely realizing westward,
But still unstoried, artless, unenhanced,
Such as she was, such as she would become.[3]

Before noon on March 30, 1867, and by virtue of two signatures on a bill of sale, the United States possessed itself of the continent's westernmost territory. Neither the two men nor the nations they represented knew the nature of that remote place. Hearsay spoke of coasts beaten by three seas, of an immense interior land of little sticks guarded by enormous mountain ranges, and of thousands of miles of intricate coastline edged with tall somber trees. A century later volumes of measurements were on shelves and ever-more-sophisticated machines were recording particles in the air, magnetic storms, wave heights, subterranean rock densities, infrared waves from land and water surfaces, photosynthesis rates and the heartbeats of diving seals. We now know much more about the country. In another century we may come to know ourselves in relation to the country. For now we can only report measurements, and that is not insignificant.

It is hard, given our own short life span and limited perspectives, to avoid describing a piece of geography as if it were isolated and never-changing. I hope not to make that mistake. Certainly Alaska is, but it also was and is becoming, at time scales ranging from hours to eons. Likewise, Alaska is an open system, not a closed one. Enormous masses of "our" ocean water actually move with gentle inevitability to us from the central Pacific and North Atlantic. Alaska's celebrated pristine air carries the chemical signatures of its earlier contact with Europe: it is only a small exaggeration to say that when Poland burps, Alaska smells garlic. Even the vigorous spring songs of our migrant birds are energized by food eaten in some distant southern wintering home.

1

Oceans of Air, Oceans of Water

As an Inuit woman once sang to Knud Rasmussen in an impromptu poem, Earth and the great weather make the real whole.[4] The complex of landscapes called Alaska is what it is only because the north-temperate Pacific lies just so to the south, and the frozen and nearly tideless Arctic Ocean to the north; only because the spinning and tilting Earth gives and takes away sunlight in a particular way, the jet stream whips sinuously high above and cold polar air meets warm oceanic air within predictable limits of unpredictability. And so this brief natural history of Alaska begins with a look at the moving masses of air and water that surround the great land.

Alaska's Climates

The meeting of polar and Pacific air masses is a key to Alaska's environmental cat's-cradle. This northwestern peninsula of North America lies where two great bands of circulating air sweep tangentially toward each other along the earth's surface. Polar air moving east and southward and oceanic air blowing east and northward meet—usually at latitudes between 52° and 60° North — and burst upward, causing low pressure storms. When polar highs are strong and thrust southwestward from their Siberian birthplace, most of Alaska is dry and—in winter—cold. When the high weakens, oceanic storms move northward from the Pacific with their wetter, warmer, blustery air. A major storm track originates near the western Aleutian Islands and moves eastward into the Gulf of Alaska, one storm often nipping the heels of another. Encountering the mountains of the Alaska and Coast ranges, each storm drops moisture and drags its weakening bulk into western Canada. In summer a secondary storm path diverges into the Bering, Chukchi and Beaufort seas, bringing its cool drizzle to western Alaska's coasts and to the Interior when the polar pressure cell weakens.

The state's only true arctic climate occurs north of the crest of the Brooks Range (Figure 1).[5] Winter—which, being a condition of the mind, has ragged edges—lasts there from early October to late May, and is persistently cold, cloudy and breezy. Residents are glad of the winter clouds, which block the heat fleeing the earth and make the cold less severe. Summer is very cool, very cloudy and very short, a green comma in a long white sentence. When open water shows in narrow leads across the ice of the Arctic Ocean in summer, fog often envelops the coast. Summer visitors splashing across its flower-strewn marshes would disagree, but geographers note this region's paltry 5 to 15 inches of annual precipitation and call it a cold desert. Day length variation is remarkable, with continuous sunlight at Point Barrow from May 10 to August 2 and continuous darkness or twilight from November 18 to January 24.

Between the Brooks and Alaska ranges lies a sprawling interior region where the seasons live their special characters vigorously. A typical day in deep winter is more than 100°F colder than a typical summer's day; recorded extremes are -80°F and 100°F, both just north of the Arctic Circle. (However, in winter when the boundary between polar and North Pacific air shifts north or south a few hundred miles, dramatic temperature changes occur. I amused myself one February day by telling a cab driver at the Houston airport that in the past four days temperatures at home in Fairbanks had risen 80°F—and it was still below freezing.) In summer there is ample light for outdoor activities all night, and in winter near the Arctic Circle you can scarcely sandwich lunch between sunrise and sunset. Although total annual precipitation is only 10 to 20 inches,

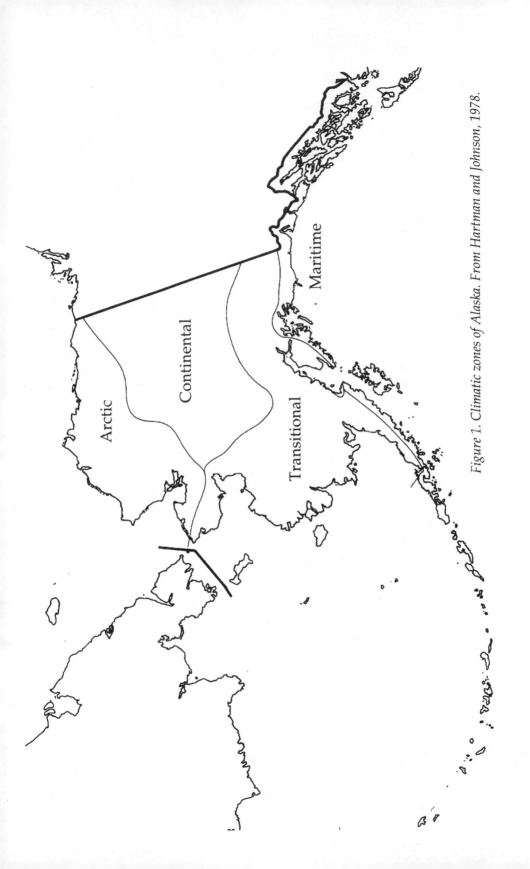

Figure 1. Climatic zones of Alaska. From Hartman and Johnson, 1978.

the boreal region's light, dry snow usually covers the ground continuously from late September to early May, commonly building to depths of 2 to 3 feet by February. The warm-season rains are uneven. Interior Alaska gardeners often have to irrigate in June and do sun-dances in August. In a memorable week in August 1967 a single rainstorm deluged Fairbanks' watershed with over 20 inches of water, almost ten times the normal total for August and twice the usual annual precipitation.

The oceans that spawn those storms create a third major climatic region of Alaska, the maritime region. Stretching from Attu Island at the beginning of tomorrow east and then south through the Panhandle to Ketchikan, this region has the wettest, windiest and warmest climate in the state. Autumn and winter storms shriek out of the Aleutians, pile up huge waves in their passage across the Gulf, and shred themselves against the towering, glacier-locked mountains rimming the entire eastern North Pacific. Summer and winter fogs are common. Wet winter snows fall thickly on mountain tops, while at sea level flakes become drops and the blanket of white wears thin and disappears in winter thaws. Annual precipitation averages 60 to 160 inches at ocean edge along the Inside Passage in southeastern Alaska and up to 240 to 320 inches on mountains facing the open North Pacific. Yet, there are breaks in the clouds, and one of the world's best kept secrets is a fine blue day in the Aleutians, the breeze slapping whitecaps against rock-black, grass-green cliffs. In the Panhandle sun-breaks occur even more often. In the memorable summers of 1968 and 1989, the Southeast basked in almost unremitting sun for weeks. The incoming tides heated on the warmed rocks and folks waded in the sea with rare pleasure. A few forest fires broke out, startling even some old-timers. The fishing town of Petersburg had to import drinking water when its normally rain-replenished reservoir emptied, and its cannery worked on short water rations. In 1989 Ketchikan's pulp mill shut down in late summer when its reservoir ran dry.

If out of about 5 billion people on Earth somewhere around a billion have heard of Alaska, it is a fair bet that most of them know only one thing about the place: it is *cold*. Wishful denials notwithstanding, we have to admit it: from the perspective of people living closer to the equator, Alaska is cold. From a resident's viewpoint, however, what matters is that some places and seasons are a lot milder than others.

Prospective gardeners and farmers, for instance, like to know when the last and first frosts occur and how many degree-days accumulate in a growing season. (Degree-days are the cumulative number of degrees above 40°F during the period when average daily temperatures exceed that minimum, and when no night frost occurs.) In Alaska most places have fewer growing degree-days than the threshold reasonable for commercial agriculture.[6] Among the districts above that threshold, the warmest is the Anchorage-Matanuska area with close to 2000 growing degree-days and a frost-free period of 110-130 days. Low-elevation areas of the central Tanana Valley (Fairbanks and nearby towns) and Yukon Basin (Fort Yukon) follow closely in degree-days but have substantially shorter growing seasons of 90-95 frost-free days.

What this means is that coldness restricts large-scale, open-air crop-growing to a few lowland areas of the Interior where long and often sunny summer days are combined with reasonably long frost-free seasons, and to the lowlands of upper Cook Inlet in the transition zone between maritime and continental climates. Many other factors, either environmental (soil type and soil moisture, for example), economic (for instance, local costs for fertilizer and machinery), or cultural (such as the lack of an agricultural tradition among Athabascan villagers in the Interior) can further restrict agricultural feasibility. On the other hand, less familiar agricultural pursuits such as reindeer and muskox herding are not bound by the usual thresholds of degree-days and growing seasons. In addition, north-adapted crop varieties have loosened present environmental bounds to some extent. We can expect use of endemic plants and newer technologies of frost-resistance to help as well. In short, climate undeniably restrains Alaskan agriculture, but not rigidly or in all areas.

For home owners and building designers, measurements of cumulative warmth or cold need to extend through the whole year. One such measurement is heating degree-days, an annual sum of degree-days below a mean daily temperature of 65°F. The contours of heating degree-days run roughly east-west over the whole state.[7] The mild-weathered southeast region has the lowest deficit below the 65°F threshold (8000 degree-days) and the Arctic coast, the highest (20,000 degree-days). To keep a room at 65°F year-round in similarly built houses, a family in Ketchikan theoretically needs to use only 40 percent as much energy as a family in Point Barrow. (Actually, the difference is even greater because Barrow is windier.)

If a climate is cold enough, the depth of ground that thaws in summer is less than the depth that freezes in winter. A band of permanently frozen ground develops just below the freeze-thaw (active) layer. Even if a local climate is now just too warm for permafrost to develop, deep-lying frozen ground may persist for centuries as a relic of ancient colder times. Alaska has relict permafrost along the southern margin of the regional permafrost zone[8] (Figure 2) and under shallow areas of the Arctic Ocean that were dry land until fairly recently. In northern Alaska the permanently frozen ground is being maintained by today's subfreezing annual mean temperatures. However, in interior Alaska the temperature regime is so near to a balance point that permafrost is absent in some sites (south-facing slopes, under rivers and lakes and in disturbed areas with bare soil) and present in others such as north-facing slopes and valley bottoms insulated with moss and peat.

The temperature and depth of permafrost vary a great deal. Patches of continually frozen ground south of the Alaska Range may be only a few yards across, very shallow, and have boundary temperatures of 32°F. On the North Slope the ground may be frozen for over 1000 feet below the active layer and may be 15°F near its upper edge. The closer to 32°F its temperature the more sensitive an area of permafrost is to small changes in annual mean air and soil temperatures.

Since about 1977 a historically unprecedented series of warmer-than-average winters has made weather experts eye the Interior's permafrost warily. If a warming pattern persists, much of the region's permafrost could disappear. Some bogs would become lakes, some ponds now dammed by plugs of frozen soil would drain, rivers would change courses, deep avalanching would rumble on steep slopes. These and other terrain effects obviously would create and destroy plant and animal communities. Just as obviously, the engineering of human settlement and resource uses

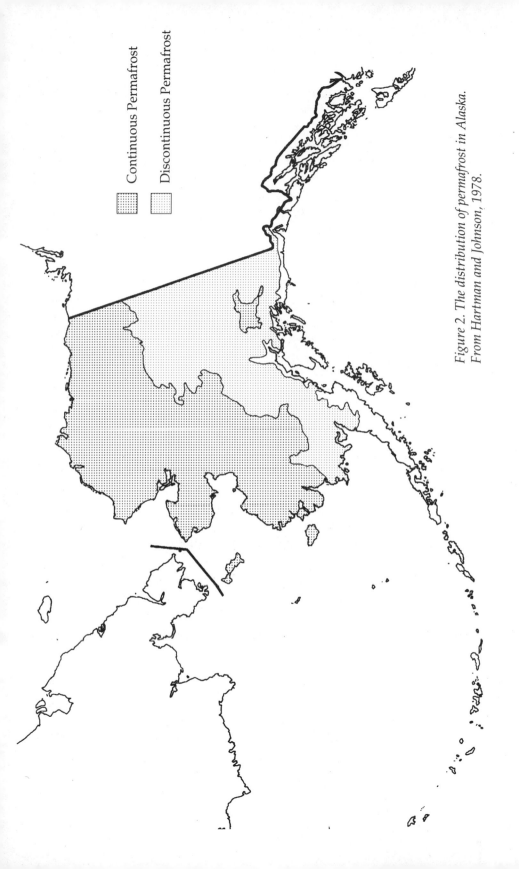

Continuous Permafrost

Discontinuous Permafrost

Figure 2. The distribution of permafrost in Alaska.
From Hartman and Johnson, 1978.

would have to be completely revised. As a wry example, the state's lucrative Prudhoe oil fields could become federal offshore property by virtue of a slight ocean level rise or subsidence of the coastal plain's permafrost—though by the time the ocean invaded the oilfields, the oil itself would most likely be long gone.

Another climate-dependent phenomenon in the North is wildfire. Only three things are necessary to create wildfire —something to burn, dry air, and a spark—and the climate and settlement of central Alaska create all three. Beginning in May with the first drying of ground cover and ending in early fall with cool and more humid weather, woods and the most thickly vegetated tundra areas provide enough fuel to support a fire if the igniting spark occurs and dry air supports its growth. Fire has an impressively long history in interior Alaska, surely extending far back into the Pleistocene glacial periods up to a million years ago when dry climate and grassland steppe vegetation created a perfect wildfire environment. More recently—for perhaps 12,000 or 13,000 years— forests, not grasslands, have provided the fuel and have been given their basic character by fire.

How often do wildfires strike in the Interior? Earlier than 1940 we have nothing to go on except bits of indirect evidence and guesswork. Often enough, certainly, to be a force in the evolution of individual plant species characteristics, and often enough to shape overall plant and animal community dynamics. Since 1940 there have been official records (not by any means complete), and study of them leads scholars to think that fires recur locally at intervals ranging from 50 to 400 years, depending on how fire-prone a particular place is.[9]

Whether fire suppression by the Bureau of Land Management and State of Alaska has lengthened these intervals isn't known. With many times more people present in modern than premodern times, and with the internal combustion engine now part of the scene, the number of forest fires started must be greater. However, the size of the biggest fires probably has dropped with the huge and costly suppression program of the last half-century.

At Cold Bay in the eastern Aleutians, people tie down their homes with cables. In Juneau winter winds funneling down the Taku River sometimes raise roofs. In mountain passes and on coasts everywhere in Alaska, the wind's fingers pluck restlessly at everything moveable. At the opposite extreme, one of the peculiarly important aspects of some Alaska airsheds is the stillness that sometimes grips them. Air temperature inversions account for many of these episodes of stagnant air, developing at any season but most common and extreme in winter. When polar highs hang over central and northern Alaska, radiative cooling under the clear skies causes a layer of intensely cold air at ground level. Unless flushed out by wind (or simply drained downhill), this cold, dense air forms a still pool with essentially no internal circulation, isolated from moving masses of warmer air a few feet above it. Inversions through which muffled Fairbanksans often wade are among the strongest and most long-lasting in the world. However, scores of places in Alaska's Interior, most of them uninhabited, are prone to air inversions. Even coastal places like Valdez and Anchorage experience inversions occasionally. As water vapor and pollutants pour from vehicles and chimneys into the temporary lakes of stagnant atmosphere, the towns within them vanish into thick air.

Oceans

Many Alaskans rise daily in sight of the sea. Many others would if the neighbor's house weren't there. Every morning thousands of eyes scan the water, watching for fish flashing above the ruffled surface, for signs of a break in the storm, for the barge bringing stores, or simply to say good morning to an old friend. Even those separated from the smell of salt water can sense the sea in the wet clouds racing overhead and in the bright salmon nosing gravel in nearby creeks. For Alaska, the sea is boundary and connector, source and sink; presence, definer, inspiration.

By geographers' count, Alaska has two oceans and three seas around it (Figure 3). The dividers are in two cases obvious even to a landsman. The Bering Strait—dry land not many millennia ago—marks the boundary between the Arctic and Pacific oceans, although currents do flow over its shallowly covered continental shelf. The volcanic, mountainous Aleutian Islands and accompanying deep trench separate the North Pacific from its big embayment, the Bering Sea. The boundary between the Chukchi and Beaufort seas is a lot less clear, lying somewhere seaward of that vague low curving coast at the northwestern tip of North America.

Nothing could be more self-evident than the fact that the oceans are a medium in motion. Only the plants attached to the bottom in intertidal and subtidal sea margins and the animals that burrow into, stick onto or crawl upon the ocean floor can remain at the same address effortlessly from one day to the next. The vast bulk of marine life lives, like birds on the wing, in motion in a mobile home. Like Alice in Wonderland, small sea creatures would have to struggle to stay in the same place; at rest they move slowly but inexorably through space with the currents.

The patterns of big ocean currents reflect persisting patterns of global air flow and the way the moving water is deflected by continents, shallows and deeps. The Japanese Current is one of those big-scale movements. Major eddies and diversions from the mainstream currents often form the primary dynamics of gulfs and regional seas, like the Alaska Current and the spring and summer northward flow of water through Bering Strait. What may be less obvious to a landsman is that these currents and the still or slowly moving waters through which they move are qualitatively different. They may be very salty or nearly fresh, warm or cold, heavy or light, nutrient-laden or nutrient-sparse, as well as relatively fast-moving or ponderous.

Patterns of biotic abundance, unsurprisingly, are related to physical and chemical sea-patterns derived partly from those large-scale water movements, partly from turbulence at a smaller scale. Most primary production (photosynthesis) occurs where water rich in nutrients is carried upward into the zone that light can penetrate. Commonly this happens at the edges of continental shelves or other deep-to-shallow transitions, or at the mouths of nearly enclosed bays, or under the paths of spring, summer and fall storm systems. The edges of moving ice can also create these same conditions of high phosphorus and nitrate levels in a well-lighted stratum.

A lot of secondary production (that is, the conversion of plant to animal tissue by grazing, and of little-animal to big-animal tissue by predation) occurs in these same places for obvious reasons. It may, however, take place further down-current because of the time

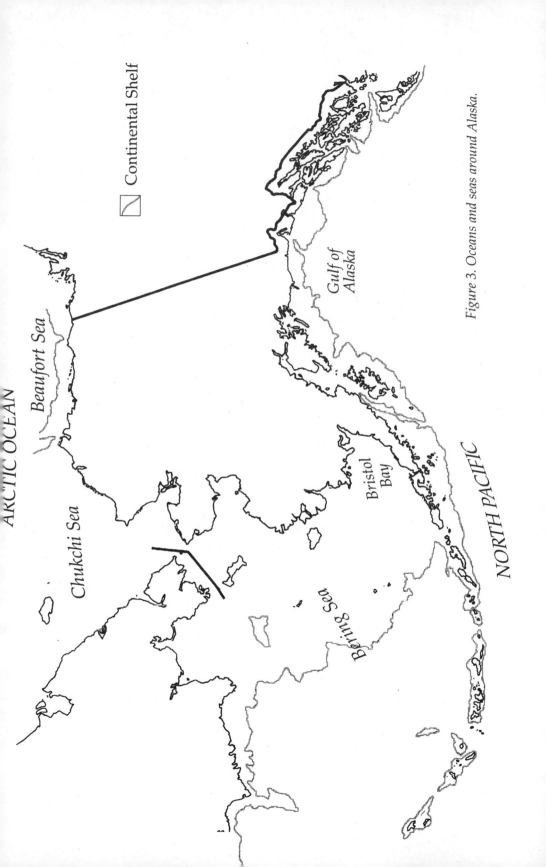

ARCTIC OCEAN

Beaufort Sea

Chukchi Sea

Continental Shelf

Gulf of
Alaska

Bristol
Bay

Bering Sea

NORTH PACIFIC

Figure 3. Oceans and seas around Alaska.

that elapses while the grazing and predation goes on. And secondary production can be quite high in eddies and embayments into which the barely mobile small zooplankton are carried or into which the vigorously swimming forms move.

Finally, a third level of biotic production is based on dead organic material consumed by detritus-eaters like sea worms and many molluscs and then transferred among interconnected predators and scavengers. This activity obviously relies on the transport of dead plants and animals, including both those freshly dead and organic debris that might be years or centuries old, by currents and gravity. The richness of the bottom-dwelling part of this scavenger-predator-decomposer food web also is influenced by whether mud, sand, gravel or rocks cover the bottom.

Light, water temperature and (in the Arctic) ice conditions are strongly seasonal in northern regions. Photosynthesis, too, is very seasonal. Within-year differences are most extreme in the Chukchi and Beaufort seas where the upper waters are dark for so long in winter. Even in the southern Gulf of Alaska, however, the winter productivity slump is very obvious, resulting mostly from excessive turbulence in surface waters due to storms. In all of Alaska's marine areas there is a strong resurgence of activity at the end of the winter, a peak fed by the combination of increasing sunlight and high nutrient levels at the surface. The spring bloom occurs in February and March in the southeastern Gulf and in April or May under the ice of the Chukchi and Bering seas. Nutrients often are exhausted by the spring activity—they are tied up in the tissues of algae and zooplankton—and hence production drops in midsummer. In late summer and fall, storm-generated turbulence renews nutrient supplies near the surface and a moderate peak in phytoplankton activity occurs before winter sets in.

North of the Bering Straits the ocean's biological wealth drops rapidly. Alaska's polar seas are, in general, poor in total living biomass and low in annual production. Some species of fish, birds and mammals, however, can be abundant if they are mobile enough to take advantage of the brief summer's photosynthesis and the zooplankton carried by currents from the Bering Sea. South of the straits and all the way to the British Columbia border (and beyond), marine life is much more abundant—commonly from 10 to 20 times as rich in carbon-fixing photosynthesis as the arctic seas. The center of this broad expanse of fairly high productivity, from Kodiak Island west through the Aleutian Islands into the southern and central Bering Sea, is probably the richest area of its size in the Northern Hemisphere. Paul Bunyan's angling counterpart, standing astride any mountain on the Alaska Peninsula or Aleutian chain, should catch a fish on every cast and backcast. The annual commercial harvests of over 7 million metric tons of fish and shellfish, nearly 10% of the total world fish harvest, and the enormous cacophony of seabirds on breeding rookeries throughout the region, are proof of this marvelous biological wealth.

Nevertheless, this wealth is neither infinite nor unvarying. Stock after stock of fish discovered in the 1950s and 1960s was overharvested in the 1960s and 1970s. Clearly, the hungers of the Orient and North America outstripped even the Bering Sea's bounty, and continuing yields are totally dependent on continuing regulation of harvests.[10] Remote as the North Pacific and Bering Sea may seem to a subway straphanger, they are no longer pristine. We changed these enormous systems before we knew what they were, and now we can never know.

Scientists are beginning to measure and interpret what fishers and native people have long known: that the abundance of marine animals varies not only from place to place, but from year to year. Colonies of sea birds are a good barometer of these quick fluxes from feast to famine because the production of bird eggs and offspring is so dependent on the food available to adults within an energetically acceptable round trip distance from the nest cliff. For the past 15 years, various bird colonies, from Prince William Sound to Cape Lisburne on the Chukchi Sea coast, have shown huge variations in breeding success from year to year. Sometimes a cluster of species feeding on similar foods (small fish near the surface, for example) would show a pattern of failed reproduction throughout a large area. At other times the failures or successes were inexplicably local, or would affect only one species.

Localized failures seem tied to local, short-term weather conditions or quirks of ice and current conditions. In one such event, foul weather for four days in April 1970, culminating in a severe storm, prevented common murres from feeding and caused a massive die-off of that species in the southeastern Bering Sea and Bristol Bay.[11] These incidents can be very important in the welfare of particular species in a certain district, but don't reveal much about how ocean chemistry, ocean temperatures, weather, competition among animals, and other factors interact to form what we picture as a dynamic marine ecosystem. The broad-scale synchronies I spoke of earlier do give us parts of that picture. For example, after pollock declined dramatically from 1969-1973 and failed to recover in the subsequent years, kittiwakes (small gulls that eat fish) and common murres—which also eat fish—nesting in large colonies in the southeast Bering Sea had very low nesting success compared to colonies in the Gulf of Alaska. At the same time, the copepod-dependent least auklets increased substantially, unhurt by the decline in pollock and perhaps even enjoying the increase in copepod numbers stemming from a drop in predation by pollock.[12] But the situation is still more complex. Sea-surface temperatures in the region naturally vary in a cyclic way, and the reproductive success of fish-eating seabirds seems to track that cycle. As well, pollock reproduction is very sensitive to changes in sea-surface temperatures.

With this reminder of the basic dynamism and changeability of marine ecosystems, I'll turn to a brief snapshot of each of Alaska's bordering seas with the zoom lens extended a bit further.

The *Gulf of Alaska* is a broadly open extension of the temperate North Pacific Ocean nestling into the curve of Alaska's steep coastal mountains. The continental shelf is narrow here over the meeting place of the massively sliding Pacific and North American plates. Deep waters lap quite close to shore. The Alaska Current, a major northward curl of the Japanese Current, sweeps north and west around the Gulf, creating a pigtail-like gyre just south of Kodiak Island. This is the current that, in the spring and summer of 1989, carried oil from the ruptured *Exxon Valdez* west more than 700 miles, past Lower Cook Inlet and Kodiak Island and onto the Alaska Peninsula. West of the gyre the Alaska Current slips along the south side of the Aleutians, sending jets of water through passes into the Bering Sea. Sea ice is rare in the main Gulf but forms yearly in the fast-cooling shallows of Prince William Sound and Upper Cook Inlet.

The entire northeast Pacific including the Gulf of Alaska has for many thousands of years supported human communities that capture its fish and shellfish. Areas near shore have yielded a protein bounty for all of that long history, and modern equipment has both intensified the nearshore harvest and carried it offshore. These fish and shellfish populations, directly or through tangled food webs, all are supported by the huge photosynthetic engine of solar energy transformation.

The millions of sea birds that plunge, dive and skim the waters of the northern and western Gulf mirror the ocean's wealth, especially the abundance of prey in surface and midwater levels. Species like albatrosses, petrels, shearwaters, and some kinds of gulls, auklets, murres, cormorants and puffins are truly marine, staying at sea all their lives except when at nests. Of these, some—including deep divers and surface feeders—breed during Alaska's summer on rocky coasts and cliffs. Something like 9 million sea birds of 26 species nest at 800 sites in the region.[13] Others, mostly far-ranging surface-skimmers like petrels and albatrosses, come to Alaska in May through October after nesting and wintering in the central and south Pacific. Another suite of bird species lives on the ocean or along its tidal edges most of the year (usually August through April or May) but breed around freshwater lakes and wetlands. This group includes most gulls and terns, several kinds of sandpipers and plovers, the phalaropes, loons and some cormorants, and all the so-called sea ducks: scoters, eiders, oldsquaw ducks and harlequins.

Marine mammals, too, exploit the Gulf's fish and other wealth. Humpbacked whales, equipped with baleen, filter zooplankton and small fish from dense concentrations. Gray whales, also using baleen plates, feed at the ocean bottom. The toothed whales, dolphins, porpoises, seals and sea lions are fish-eaters, each pursuing its unique strategy of food-gathering opportunism or specialization.

Shaped like a mammoth's tusk, the Alaska Peninsula juts southwestward out of Alaska's bulk, then continues in a thousand-mile arc of islands toward Russia's Kamchatka Peninsula. North of this curve of green-and-black islands is the *Bering Sea*. Enclosed by the treeless west coast of Alaska, the Aleutians, and the Siberian coast, the Bering Sea receives water from two oceans and several major rivers including the Yukon and Kuskokwim. The northern half of the Bering Sea lies over a flat, shallowly covered plain, part of the continental shelf connecting Asia and North America. Much of the Bering Sea floor is covered with sands and muds brought from the Alaskan Interior by the Yukon and Kuskokwim rivers. South of a wriggling shelf edge trending northwest from the easternmost Aleutians toward Siberia's Cape Navarin, the Bering Sea is much deeper. Ice forms across Bering Strait in October, and a little later in the quiet embayment of Norton Sound. From December through March drift ice covers 10-70% of the sea over the continental shelf.

The broad expanse of shallow water, the mixing of waters of different origins, and the winter buildup of unused nutrients characteristic of northern seas make the Bering Sea one of the most productive in the world.[14] A current (Anadyr Water) that flows deeply along the continental shelf and eventually brings nutrients toward the surface in the northern Bering Sea is especially important. It also transports tons of zooplankton daily across the shelf break into the northern Bering Sea. The Bering Sea contains some of the largest marine mammal, seabird, fish and crab populations anywhere. The most productive commercial fishing area in the Sea is along the shelf break and on the continental shelf in the southeast quadrant.

Ice is an extremely important factor in the distribution of Bering Sea birds from late fall until late spring. The seasonally moving pack ice front has an enthusiastic following. Here are heavy concentrations of gulls and murres, sometimes in flocks of 25,000 individuals, all feeding primarily on pollock, which in turn are attracted to amphipods and euphausiids that occur in clouds at the ice front. Within the dense and extensive pack ice, however, birds are sparse. Open areas (polynyas) occurring leeward of large islands are refugia for diving birds in early winter, but only ivory gulls and black guillemots are adapted to a truly ice-dominated sea.

The large bowhead whale winters in the Bering Sea, and gray whales from Mexico and California summer there. Beluga (white) whales are common near the shore when salmon concentrate in estuaries. Fur seals, of which between 800,000 and 900,000 call the Pribilof Islands home, travel very widely for fish. Walrus follow the north- and south-moving ice pack, being abundant in winter in the central and southern Bering Sea. They are the only large animal feeding heavily on the extensive clam populations of the region.

Seen from a satellite high over the North Pole, the *Chukchi Sea* is a gulf of the Arctic Ocean in the angle between Alaska and Siberia. It is an odd gulf, because it is pinched but not wholly blocked by the Bering Strait on the south, so that Arctic Ocean and Pacific Ocean waters can join. The Chukchi Sea's neighboring *Beaufort Sea* seems, from the same height, to be without any distinguishing boundaries; it is merely the part of the Arctic Ocean west of the Canadian Arctic Archipelago. The Chukchi Sea shallowly covers an extremely flat plain, the Chukchi Shelf, part of the continental shelf connecting Asia and North America. North of Point Barrow a deep canyon plunges down to the depths of an abyssal basin, the Canada Basin, between 8,000 and 12,000 feet deep. The Canada Basin comes quite close to shore north of Alaska's Arctic Slope, leaving a narrow but very shallow continental shelf at the rim of the Beaufort Sea. Water circulation in the entire region is dominated by a north- and then east-flowing current carrying Bering Sea water and coastal runoff close by the Alaska shoreline, and a westward sweep of water farther offshore, originating in the Atlantic Ocean and caught in the enormous circulating current of the Polar Basin.

All this expanse is sheathed with ice most of the year, and some of it is covered all year. Shore-fast ice forms early in winter and builds up 6 to 8 feet in thickness until, early in spring, it slowly rots and crumbles away from the warming shore. Pack ice drifts around the Arctic Ocean all year, breaking up in summer and solidifying and expanding rapidly in winter. Pack ice and landfast ice are jammed together throughout winter and early spring, forming a shear zone of crumpled and contorted ridges. Openings appear in May and June as warmer Bering Sea water pushes north along the Alaska coast beyond Bering Strait. The opening of this major shore lead is a biological as well as physical red-letter day in the arctic calendar as whales, seals and millions of birds follow its erratic northward progress.

Coastal waters in the Beaufort Sea have very limited amounts of nitrogen and phosphorus. Coupled with the extremely long period of dim light or darkness, these limitations reduce annual primary production to low levels. An early spring surge of photosynthesis (when "spring" utterly fails to describe what the weather is like above the ice!) takes place just under the clear ice, when light increases and algae capitalize on the overwinter buildup of inorganic nutrients. Zooplankton graze these phytoplankton. Copepods, amphipods, shrimps and euphausiids are common pelagic crustaceans in the Arctic, sustaining the ecologically important arctic cod populations as well as summering belukha and bowhead whales, seals and seabirds.

In summer the two most abundant marine fish (arctic cod and fourhorn sculpin) and the three most numerous anadromous fish (arctic charr, arctic cisco, least cisco) of the Beaufort Sea are much more common nearshore than offshore. They travel widely along the coast, feeding on crustaceans and small fish. Little is known about where these fish go in winter. The cod presumably move offshore, and apparently spawn under the ocean ice in midwinter. The anadromous fish probably move into rivers and lakes, although some spend winter months in brackish waters of the Colville River delta.

A few marine mammals are well adapted to year-round or seasonal life in arctic waters. Ringed seals, found mostly in landfast ice in winter and spring, feed on cod and medium-sized crustaceans and themselves are a basic food of polar bears. Bearded seals live in pack ice farther offshore and seem to eat a variety of invertebrates on the bottom. The endangered bowhead whales, which winter in the Bering Sea, move along the shore lead in May and June into waters of the Beaufort Sea, feeding on crustaceans.

Birds are essentially absent from the Chukchi and Beaufort seas from October to April, although a few species winter not far away in the Bering Sea. The spring influx totals millions of birds, including loons, geese, phalaropes and sandpipers that use tundra freshwaters in summer either in Alaska or northern Canada; eiders, oldsquaw ducks, gulls and terns that forage extensively in nearshore waters; and colony-nesting seabirds (murres, cormorants, puffins and kittiwakes). Cape Thompson and Cape Lisburne, where the sea meets the western end of the Brooks Range, form the only two major bird cliffs in the entire Chukchi/Beaufort system.

Alaska's bordering seas, then, are distinctly northern. But what does that mean? Certainly it implies a sharp pattern of seasonality: seasons of storm and relative calm, alternating seasons of intense light and darkness, seasons of vibrant life activity and times of low (but never absent) energy and nutrient exchange. Northernness also implies a pervasiveness of cold, but in fact cold temperatures are less dominant as physical or biological controls than you might expect. Life's activities can go on as long as the sea's water is liquid. When the sea surface is frozen, of course, the ice and its snow load dramatically change conditions of life in the strata below it.

Northernness, in the case of northern hemisphere oceans, does not necessarily imply biological impoverishment. The cold northern Gulf of Alaska and the colder Bering Sea are exuberant with life. Beginning with one-celled photosynthesizers in the upper few feet of those waters and curving around to the detritus feeders of the ocean bottom, living on recent or ancient organic fragments, those regions have an abundance of life hard to match anywhere in the world's oceans. It is generally true, however, that north of Bering Strait the polar seas are severe environments to which relatively few species are adapted, and in which the total annual capture and transference of solar energy is low. In either the biologically generous or the biologically stingy regions, physical and chemical conditions and the communities of plants and animals responsive to them can change dramatically from year to year.

2

The Land and Flowing Waters

It is as if a god's left hand drew the strands of all of western America into a curving convergence to the northwest and then plunged them into the Arctic and Pacific oceans.

Look at a map of the Great Plains. Broadest from Minnesota to Montana, the Plains undulate northward through boreal Canada, narrowing continuously until, after pinching out briefly in northern Yukon, they become Alaska's Arctic Coastal Plain. Bunchgrass in the heat of west Texas has become cottongrass tussocks in the cold of northern Alaska.

Then look at the western cordillera of the contiguous United States, the Rocky Mountains, the fabulous wall that for thousands of years divided Indian tribes and for centuries enticed Euro-American explorers. North the ranges run for two thousand miles, receive the wild arc of the MacKenzie Mountains, then, turning west and now bearing the name of another explorer, Alfred Brooks, they divide Alaska's true Arctic from the quasi-arctic and boreal remainder. Behind the Rockies—again in the West—is the Great Basin and a jumble of plains and short mountain ranges geologists call the Intermountain Plateaus. In Alaska these are the ancient, unglaciated uplands and valleys of the Interior: the Yukon Valley, the Kuskokwim Valley, the Tanana Uplands, the Kuskokwim

Mountains and a dozen other ranges and valleys full of mystery even to most Alaskans. Finally there is the Pacific Mountain System that in the western U.S. and British Columbia is known loosely as the Cascade and Coast ranges and in Alaska becomes a breathtaking sweep of jagged, glaciated mountains including the Alaska, St. Elias, Alaska Coast, Chugach, Kenai and Aleutian ranges.

This whole mountain complex is geologically active. Earthquakes are very common. The mountains around the Gulf are part of the famous Ring of Fire in the north and west Pacific. The Aleutian mountains contain 57 volcanoes, of which at least 30 are active. Among the numerous mountain glaciers, one or another of them is almost always in spectacular advance or retreat. In the early 1980s the retreating Columbia Glacier left hunks of ice behind that endangered the oil tankers visiting the Port of Valdez in Prince William Sound. (Curving to avoid ice, the *Exxon Valdez* in March 1989 ran aground on Bligh Reef, becoming yet another spectacular statistic in the long history of spilled oil.) In the summer of 1986 the Hubbard Glacier suddenly advanced at rates that reached over 100 feet per day, creating a big lake and an enormous prow wave of gravel not far from the town of Yakutat on the shore of the eastern Gulf of Alaska. The lake drained a few months later, but in 1989 the cycle started afresh. And in the summer of 1988, scientists announced that the Taku Glacier, advancing under the weight of record snowfalls at the rate of 600 feet per year, would in only 6-8 years totally block the Taku River and its salmon run.

In describing what Alaska's landscape is like, everyone looks first to the mountains. Yet the lowlands, too, are fascinating. The plains of the Arctic with their frost-patterned ground and wind-oriented lakes; the sand dunes and lava beds that surprise travelers in western Alaska; the oxbow lakes, muskegs and serpentine rivers of the Interior; the braided glacial streams and enormous outwash fans at the feet of mountains all over southcentral Alaska; and the morainal hieroglyphics wherever Pleistocene glaciers grumbled, all tell their own tale of force and counterforce, building and breakdown.

Mantles of youthful soil cover nearly all these landscapes. They are phenomenally mobile and active. The riven outer rocks of mountains constantly tumble, slide or creep downslope, singly or in massive screes, cracked by frost as they move and chemically weathered into new soil particles. Glaciers grind rocks into an exceedingly fine flour which, carried by water or wind, is deposited in deltas and floodplains and on hillsides. Freeze-thaw cycles, endlessly repeated, churn the ground and sort soil particles by size. Streams cut, carry and dump. Lava and ash flows start new soil-creating processes on top of earlier ones.

In such a dynamic landscape, the variety of local soil conditions is almost infinite. Each soil has its own set of capabilities and limitations for growing food, storing water, recharging aquifers, supporting roads and structures, nurturing wildlife and meeting many other needs. Speaking very generally, however, there are three broad groups of soils in Alaska. The most widespread group, consisting of any texture or parent material, is defined simply by the fact that it is raw and unsettled, essentially unchanged by true soil-forming processes, with no organic matter incorporated within it. Next most common are wet, organic soils where waterlogged peaty accumulations from two to scores of feet thick lie in shallow depressions. Over most of Alaska these soils are perennially frozen, but in southern (especially southeastern) Alaska they are warmer. Bogs and muskegs form on these soils. A third soil series, better drained, loamy, or silty and often wind-deposited, is common in valleys in central Alaska. Some organic matter is incorporated into the upper few inches, and these soils often support good tree growth and crops.

North of the Brooks Range vast areas have deeply frozen soils that are either raw and without humus, or waterlogged peat. Upper layers thaw to depths of one or two feet in summer. Even land surfaces undisturbed by flooding or glaciation for over 5000 years usually show little evidence of true soil formation.

In central Alaska the highland soils are a confused jumble of depths and textures. Underlain by permafrost, they have formed in place from weathered bedrock, drifting slowly downhill from gravity's inexorable pull. The low-lying ground consists mainly of river-borne and wind-blown sediments dotted with peat- and muck-filled basins. Most of the lowland soils are waterlogged in summer and contain permafrost. From a farmer's or forester's viewpoint, the best soils are on well-drained natural levees, outwash plains and fans, and on certain hill slopes with deep silt, good drainage and the right tilt to catch the spring and summer sun.

The Brooks Range, Alaska Range, Wrangell Mountains, Chugach Mountains and other ranges of southcentral Alaska have little or no developed soil. Lower slopes of narrow mountain valleys and creekside strips have accumulations of fine materials that might be called soils, but they are, at best, extremely immature. In the Copper River plateau and Cook Inlet lowlands, gravelly morainal areas, wet, frost-churned soils on gentle slopes, deposits of silt and clay, and saturated organic soils are intricately interspersed.

Along the Alaska Peninsula and on Kodiak Island, soils are predominantly volcanic. Ashy soils in this complex are very light when dry—but they are hardly ever dry. Some support forest growth, but most do not, although a natural invasion of forests is occurring in many places in this region in lagging response to earlier climatic warming.

Most of southeast Alaska is rough and mountainous with gravelly, immature and shallow soils overlain with forest litter. Transported gravels, sands and silts at creek mouths, often flooded by fresh or tidal waters, form wet soils supporting marsh vegetation. The muskegs that are so characteristic of gentle slopes and flatlands in this region consist of deep, waterlogged deposits of moss, peat and muck.

Living Communities on the Land

Take a bit of rock. Crack it with frost. Let the sun heat it and water bathe it. Let it stand in parades of centuries. To this mineral mixture add the miracle of seeds. Fibrous hands clench down, green shoots spear skyward. And then a busy elemental traffic cycles from earth to heaven, quickness to death, and back again. First a plant, then vegetation, then animals to evolve in community with them, the whole becoming a pulsing system of life. Each of Alaska's myriad mixes of land form, soil and climate becomes a distinct possibility that some set of adapted and available plants and animals will turn into the reality that we see when we walk the countryside.

There are two great groups of terrestrial plant and animal communities in Alaska: the forest and the treeless tundra. Just as a superficial look at forests shows that the spruce-dominated boreal forests differ from the bigger, wetter needleleaf forests of the Gulf coast, so does a brief study suggest major differences between true arctic tundras of long polar history and subarctic and alpine tundras inheriting many southern high-altitude species. Each of these four can be subdivided further according to the dominance of grass, moss, shrub or tree life forms, and along other dimensions such as moisture gradients (see Figure 4).[15]

Arctic

In one way or another, temperature molds the lives of arctic animals and plants more than anything else. Nothing correlates better with arctic plant diversity than midsummer air temperatures, and soil temperature directly or indirectly determines plant growth rates.

There are many remarkable adaptations by which tundra plants cope with low temperatures. The result is that in early summer, daily growth rates in tundra grasses and sedges may be as high as among grasses in the American prairies. However, phosphorus and nitrogen are scarce in arctic ecosystems. Facing nutrient scarcities, tundra plants, nearly all perennials, limit the annual increment of tissue and maintain high nutrient reserves. These reserves, mostly below ground in winter, are transported upward from roots early in summer to support shoot growth. Less than two months afterward, the above-ground parts stop growing and nutrients are moved back below ground.

About 80 percent of the tissue grown each year eventually returns to the soil when arctic plants or plant parts die. This material is eaten by detritus-decomposing microorganisms and fungi. These are fed upon by mites, fly larvae, springtails and enchytraed worms. Many tundra birds depend almost entirely on an abundance of these invertebrates and their insect predators for food, including shorebirds in wet tundra and longspurs, pipits and upland-dwelling shorebirds like golden plovers in drier sites. Most of the remaining 20 percent of net plant growth is eaten by rodents and large mammals, as there are relatively few plant-eating insects in tundra ecosystems.

From the standpoint of wildlife, the most striking characteristic of Alaska's Arctic is the extreme seasonal change in animal numbers and diversity. The birds in their millions come late in May or in June and leave again in two or three months. Grizzlies, marmots and ground squirrels submerge into sleeping chambers in fall. Most caribou desert the Arctic in winter, different herds moving

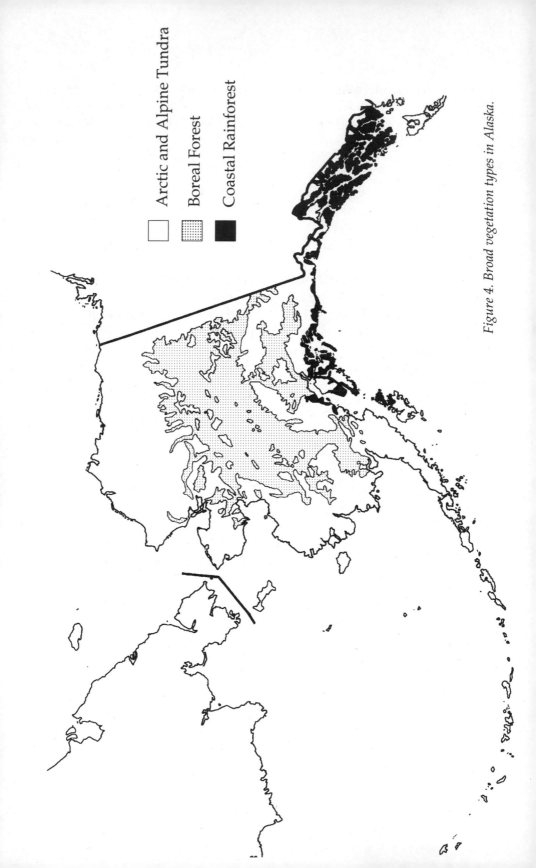

Arctic and Alpine Tundra

Boreal Forest

Coastal Rainforest

Figure 4. Broad vegetation types in Alaska.

southwest and southeast across opposite ends of the Brooks Range into the shrub and forest lands of Alaska and Yukon. Freshwater fish draw back into sparse refugia in deep lakes, springs and isolated deep pools in the ice-locked rivers.

Central and Southcentral

This region (really two, which I have brought together for convenience and conciseness) has forest as well as alpine and coastal tundra environments, and its living landscape is more complex.

Rimming Kotzebue Sound, covering most of the Seward Peninsula, and throughout the Yukon-Kuskokwim Delta is a series of tundra plant groupings much like those of the Arctic. Water-soaked sedge-meadow tundra prevails on the Delta as well as the northwest corner of the Seward Peninsula and deep in eastern Kotzebue Sound. Tussock and heath-scrub tundras are common in hilly areas. Away from the coast the central and southcentral Alaska highlands are covered with alpine tundra. Sparse perennial herbs, tiny grasses and sedges, and creeping shrubs prevail on dry ridges and upper slopes. On warmer slopes and on gravelly plateaus close to timberline, scrubby willows, dwarf birch, blueberries and other small upright shrubs and perennials are abundant. At timberline (usually 2,000-3,000 feet above sea level in this region), the shrubs become quite tall and form thickets in openings of the subalpine forest. Tussocky tundra waylays hikers on lower slopes where permafrost underlies the summer-wet surface soils.

There are marshes and muskegs below timberline, but woodland dominates the landscape. Only six species of trees are found in the sprawling Interior forest, but their combinations of species, size and spacing seem endless. Pure stands of black spruce cover extensive areas of cold, wet soils. Pure white spruce stands occupy many warmer, well-drained soils that have not been burned or cleared for one or two centuries. Mixed forests of these two species are even more abundant. Paper birch intermixes with the conifers on a variety of soils and sometimes covers the landscape after wildfires. On warm and drier soils, trembling aspen occurs either mixed with birch or spruce or as a single-species stand after fire. Balsam poplar is widely distributed along river floodplains on gravels and silts, and inexplicably, on some warm hillsides too. Tamarack grows in wet bogs along with black spruce.

Under the trees are shrubs, herbs, mosses and lichens that are important in defining the local biological community. Shrubs are virtually absent in dense evergreen woods. Deciduous and mixed conifer-deciduous stands always have shrubs such as rose, soap-berry, currants, Labrador tea, viburnum, crowberry and cranberry. Mosses and lichens grow on moist forest sites; those often over-looked primitive plants may have as great a mass and annually produce as much tissue as the trees towering above them.

In the great river lowlands, the forest cover is interrupted by stretches of muskeg and pond-dotted marshes. Muskegs may be entirely treeless but more often support sparsely scattered and very stunted black spruce and tamarack. Low shrubs are dense above a continuous carpet of moss: Labrador tea, leatherleaf, willow, dwarf birch and blueberries are common. The wettest areas have floating bogs of sedge, sphagnum and semiaquatic herbs. The marshes have less shrubbery and more grass or sedge than the muskegs. Around naturally drained lakes, bluejoint grass forms broad mead-ows that last until destroyed by permanent flooding or invaded by trees and willows.

Trees and warmer summers add new dimensions to the array of wildlife in central and southcentral (as compared to arctic) Alaska. Summering birds include many tall-shrub and tree-dependent species like warblers and thrushes. Mammals bound to forest or thicket environments (beaver, porcupine, red and flying squirrels) live in the region. Forested habitats in this part of Alaska have fewer species and sparser populations of summering birds than counter-part habitats of southern boreal and north temperate regions, but tall shrub communities have more.[16] Moose replace caribou as the most widespread ungulate, although caribou are in most of the region's main highland areas. Warmer summer air temperatures and the addition of tall shrubs and trees to the vegetation create niches for many herbivorous insects, their insect predators, and insectivorous birds. Some of the plant-eating insects (spruce beetle, spear-marked black moth, aspen tortrix, bark beetle) reach high enough numbers occasionally to cause obvious damage to interior Alaska forests. Recently, scientists have begun to think that plant-eaters like insects, moose and hares may, with their foraging, trigger a series of responses that reshape the forest itself.[17] Their browsing has put pressure on plants to develop chemical defenses that reduce the edibility and digestibility of leaves, buds and twigs. These chemicals fall with the leaves and litter to the ground and inhibit decomposition by all the microscopic life in soil. Litter builds up, and, shading soil, cools it off. This process will slow the release of nutrients and reduce productivity, and may even encourage permafrost formation.

Most birds leave interior Alaska in August or September. Still, over 25 species stay year-round. In southcentral Alaska departures take place a few weeks later, and more species overwinter locally. They are gleaners of seeds, berries, buds and dormant insects; scavengers; and predators of small mammals. All of the mammals (except possibly the uncommon little brown bat) stay all winter, too, though several hibernate, and a number of mice and shrews winter almost entirely under the insulating snow. The few boreal herbivores (voles, hares, muskrats, beaver, squirrels, moose, caribou and Dall sheep) support mammalian predators like weasels, marten, red fox, lynx, wolverines and wolves.

South of the Alaska Range the effects of higher average annual precipitation and temperature are visible. Shrublands are much more extensive, the high snow and rainfall on mountain slopes resulting in thick growths of tall willow, alder, mountain ash, currants, salmonberry and devilsclub. In the forest black cottonwood is added as a riverbank species. From northern Kodiak Island eastward the coastal forest changes dramatically from boreal to north temperate, being dominated by Sitka spruce, hemlocks and cedars instead of white and black spruce.

The special character of southcentral Alaska, for me, is at the marvellous border (or, better, communication zone) between sea and upland. Our family has a cabin near Homer, at the edge of Kachemak Bay. Here the man-tall bluejoint grass flows across the bold terraces, then tumbles over the rim of clay bluffs in great clods, a living green glacier calving into the sea. Mice burrow industriously deep in the layers of old wild hay. Above them, sparrows cling slant-legged onto slender raspberry canes. Flares of fireweed brighten the meadow in August, subsiding smokily in September.

On glowing summer nights tons of dew are caught in twined nets of grass, soaking children who by day hide like gleeful secrets in the meadow's heart.

When the sods slide down to the drift line of storm tides, they are swallowed by the sea to become elements of a different kind of life. Figuratively, bluejoint is eelgrass, moss is rockweed, loam is black salt mud. Mice are sculpins, squirrels are Irish lords, and the wind itself is reborn in the slow gyres of the salty bay. While crowds of calling birds ebb and flow annually over the meadows, under the waves tides of salmon mark the calendar.

The wildlife of southcentral Alaska mixes northern with obvious southern elements. Mountain goats, for example, which moved into Alaska from the south after the Wisconsin glaciation, are in all southcentral Alaskan mountain ranges with relatively high snowfall. Moose are common, but caribou are restricted to northern and western fringes of the region where snow is not excessively deep. Introduced black-tailed deer inhabit some islands along the Gulf. Among birds, species like great blue heron, rufous hummingbird, Steller's jay, northwestern crow, magpie, song sparrow, siskin and red-breasted nuthatch betray the region's temperate affinities.

The stormy waters between islands in the Aleutian chain have barred essentially all land mammals except for those brought in by people: rats, red and arctic foxes, caribou, bison and domestic livestock. Land birds are few in the Aleutian Islands, partly because foxes and rats brought to the Islands by Russians and Americans scoured out the vulnerable ground-nesters.

Southeast

Here, the vegetation is more like that of coastal British Columbia and the Pacific Northwest than like the rest of Alaska. There is no arctic tundra and the alpine tundras on steep, often ice-capped mountains are very patchy, interrupted by glaciers and glacial rubble, snowfields, cliffs and forested valleys. On the drier alpine ridges, mat and cushion plants are common. In moist places closer to timberline (2,500 to 3,000 feet above sea level), and in openings within the subalpine forest, tall grass and flowers give the slopes a lush emerald color in summer. Alder thickets are common at timberline and on avalanche tracks.

Southeast Alaska forests are dense, evergreen, shadowed, mossy, cool and wet. In southern areas of the region, species like red cedar, western yew and Pacific silver fir signal temperate influences. Riverbanks often have groves of black cottonwood and red alder on old gravel deposits. The forest understory—blueberries, copperbush, devilsclub, salal and other shrubs—contains many plants abundant in British Columbia, Washington and Oregon. The muskegs have scattered lodgepole pine and mountain hemlock in a matrix of low heath and sedges, the whole struggling through a thick mat of watersoaked sphagnum and dotted with small sudden pools. Grass-herb meadows occur at tidewater where creeks and rivers have piled sand and silt.

Southeast Alaska has few tundra animals (deer, bears and mountain goats in summer, along with ptarmigan, water pipits and rosy finches). Among the land birds, a higher proportion than elsewhere in Alaska is resident all year. Geologic time is an important dimension for wildlife in the region, especially land mammals. The region was ice-covered until 6,000 to 10,000 years ago, and mountain and water barriers have slowed and channelled the recolonization process. Mountain goats, red squirrels and marten were unable to reach Baranof or Chichagof islands (two of the largest and most mountainous in the Alexander Archipelago) until brought there recently by people. Deer have reached all suitable habitats, but wolves, which followed the deer, have not reached Admiralty, Baranof or Chichagof islands. Neither have black bears, whose entry was from the south. The big brown bear, on the other hand, is found only on those three islands and on northern mainland areas. Among small mammals, peculiar distribution patterns abound which relate to dispersal barriers more than to habitat quality.[18]

Rich Lands, Poor Lands

On land as in the sea, green plants do nearly all of the basic work of combining solar energy with chemical elements to manufacture living tissue (autotrophic bacteria, unlocking energy from sulfur molecules, contribute to that work too). Every other strand and junction in the ecosystem web is suspended from that main cable, including, of course, our own life and endeavor. For that reason it matters a great deal to us what level of primary production our home landscape can achieve, what causes variation from place to place and time to time, and through what channels of animal life this photosynthetic produce naturally moves.

(A note on the measurement of plant productivity might be useful. The most common, and the one I'll use, is the amount of carbon—a basic component of all living cells—"fixed" in the complex molecules of sugars, starches and proteins through photosynthesis by plants in one square meter of ground in a particular place in one year. The shorthand notation is (for example) 40 grams C/m^2/yr. Annual growth actually creates much more living tissue, but some of the year's yield is counteracted by the death of older plant parts and even more is translocated to roots. It is mostly a convenient tradition, then, that scientists use net production of above-ground tissue—excluding lichens, whose growth is so hard to measure—as an index of total primary productivity.)

If you sampled the whole spectrum of treeless tundras in Alaska you would find tremendous variation in productivity from one place to another, roughly from 1 to 100 g C/m^2/yr. High, dry ridges where plants are tiny and widely scattered would be at the low end of the scale, and wet coastal sedge meadows and moist shrubby tundras near alpine timberlines would be at the high end. Although cold air temperatures make tundras in general among the least productive kinds of vegetation, soil moisture variations have more influence on the production of different sites within the whole tundra biome. It is common to find places where primary productivity is 5 times that of other tundra habitats a few meters away.[19]

In terms of animal life in the tundra, invertebrates outweigh all of the birds and mammals that people tend to think of first. Most plant production in the long run is channelled into nematodes, mites, springtails, other insects and other less numerous invertebrates. However, as I have mentioned, only a fraction of the invertebrates are feeding on current-year plant tissue. By far the majority live on decaying plant fragments that may be decades or hundreds of years old.

Among tundra herbivores small mammals—voles, lemmings and ground squirrels—often have a bigger total biomass (as well as faster reproduction) than the larger grazing mammals. Small mammals alternate between virtual absence and teeming abundance on the tundra, reaching peaks every 3 to 6 years. The irruptions of lemmings in Scandinavia, Siberia and parts of arctic North America are legendary. At Barrow, careful records over two decades showed that lemmings reached peak densities of 40 to 80 per acre four times, each time tumbling within a few months to near zero on study plots.

At each major irruption lemmings destroy a substantial part of the grassy and mossy tundra vegetation. Their abundance means boom times for foxes, weasels, owls, jaegers and other predators. Despite the obvious damage done to vegetation in peak years, this grazing and burrowing activity increases plant growth on average because grazing increases decomposition rates, releasing nutrients into the soil.[20]

Forests are much more strongly three-dimensional than tundras; the dominant location of photosynthesis varies from tree top to ground level. In a mature dense spruce forest with few shrubs or herbs under the trees, photosynthesis is concentrated in tree crowns. In woodlands and forest bogs where trees are scattered, primary production takes place mostly in moss and shrub layers.

Overall, primary production seems to range from about 30 to 230 $C/m^2/yr$ in Alaska's Interior forests.[21] No other factor causes as much variation in productivity among different sites as soil temperature. Cold soils are least favorable because essential nutrients remain locked in undecayed organic matter. Highest productivity is in streamside tall shrubs and in young forests (regrowing, for example, after a wildfire) on warm slopes. One study showed that a black spruce stand on a permafrost-free upland site had 80% greater annual production than a similar stand on thawed muskeg soils over permafrost. Generally speaking, boreal forests are less productive after a long fire-free period when moss thickens and soils get colder, than just after fire.

There have been fewer measurements of photosynthesis activity in Alaska's boreal forests than in tundras, and even fewer in the rainforests of southeast Alaska. What estimates can be made in that coastal region are mostly extrapolations from information on yearly increments of wood on the trunks of trees. Most likely, primary production in various southeast Alaska forests averages 200 to 300 g $C/m^2/yr$, somewhat higher than forests of the Interior.[22]

In overall pattern these measurements reflect what common sense would suggest. In Alaska's land ecosystems biological productivity gets higher from north to south and from high to low elevations. Almost all of Alaska's plant communities manufacture less living tissue each year than the general run of temperate and tropical forests, by factors ranging from less than 2 to 10 or 20— ignoring extremely severe and barren arctic sites. Variation from site to site is strong in Alaska, more so in the tundra than in forested areas. Cold air sets growing season boundaries and in that way restricts annual production, but soil moisture and soil temperature explain more of the place-to-place variation. Animal biomass probably follows patterns of variable plant productivity, though solid information is scarce (only two vertebrate groups seem to have been studied with this question in mind: small mammals in tundra ecosystems and birds in Interior forests and shrublands). However, this last rule probably has many exceptions: for instance, detritus-based food webs may not reflect current annual photosynthesis very closely at all because they can use ancient carbon in peat.

Flowing Water, Still Water

The North holds many surprises for the newcomer who loves the lively water of ponds, lakes and streams. Here, some of the driest soil and scarcest surface water can occur in the wettest climates, and a countryside so full of lakes that it is impossible to walk dryshod ten minutes in a straight line occurs in a region with less rain than many deserts. Thousands of lakes perch on frozen ground. There are lakes in volcanic calderas and lakes in land-locked fiords. Thousands of miles of streams swell to near flood levels in hot, dry weather. Countless ponds and streams turn to solid ice in winter.

Taken altogether, the ponds, lakes, wetlands and streams that map the return of snow and rain to the sea are extraordinarily important in the life of the North. People cluster around them for the food they provide, for the utility of water itself, for the paths they offer through trackless hinterland—and, I often think, just for the sheer joy and companionship of glinting, moving, living water. Streams are the ancient routes of trade, exploration and commerce, overtaken now by aircraft and roads in this era when everything seems to be needed urgently, but still quietly carrying the traffic of more personal, more deliberate endeavors. And they serve a hundred critical functions in the maintenance and evolution of northern environments. They host the young of salmon and other anadromous fish in their hundreds of millions, nurturing eggs, fry, fingerlings and juveniles that likely could not find the necessary combination of oxygen, food and protection from predators if spawned in the ocean. They collect nutrients from the land and redistribute them in floodplains and deltas, often forming the richest of northern land environments. They make molehills of mountains, given time enough, but they also carve canyons and chasms through plain and peneplain. As they collect into massive rivers and finally meet the sea, their waters change the very character of the ocean itself along the rim of the continent, and bring thousands of tons of organic matter out of the uplands to contribute to food chains at the ocean's floor.

Arctic Alaska has less than 8 inches of rain and snow in an average year but still has ample fresh water. The thousands of ponds on the Arctic Coastal Plain are thaw ponds, the result of meltwater collecting in shallow tundra depressions and, concentrating solar energy, thawing frozen soil beneath. Most thaw ponds are only 3 to 15 feet deep. Those less than 6 feet deep usually freeze to the bottom in winter. Persistent summer winds from the northwest push waves against the windward end, thawing and crumbling the frozen soil and sending the lake on a slow southeasterly swim across the tundra. There are a few lakes in the Brooks Range and its northern foothills. They are deep (15 to over 100 feet) and lie in glacier-gouged valleys. They usually are free of ice only in July, August and part of September.

By midwinter most arctic streams, except those with strong springs, have no surface flow, and usually no subterranean flow either. Even the largest stream in the Alaskan Arctic, the Colville River, scarcely flows after October. That river and other major streams often have isolated pools under the ice near their mouths, important for overwintering fish.

South of the steep and lonely crest of the Brooks Range almost the whole Interior is drained by the Yukon or Kuskokwim rivers and their tributaries. The Yukon's northside tributaries receive no glacial outflow and tend to be quite clear except during spring breakup and summer rains. The small streams draining the hilly plateaus of interior Alaska are shallow and clear, deepening only when they cut into the alluvium of major valleys. Tributaries of the Yukon originating in the Alaska Range often carry glacial flour in summer. Those tiny particles added to silt from crumbling banks keep the rivers turbid until freeze-up early in October. Big streams in the Interior often flood in spring due to ice jams or rapid snow melt. Rain-caused summer floods are infrequent, but they can be severe. Glacier-fed streams may flood in their middle reaches when hot weather melts glacial ice rapidly.

West and north of the Alaska Range, where an enormous volume of sand and gravel outwash forms a plain gently sloping toward the Tanana and Yukon rivers, lakes of moderate size are found in glacier-dug depressions. The highlands of the region have very few lakes. Low-lying ground, however, has thousands of ponds and lakes of varying origins and characteristics. Most, by far, are in wide basins along major rivers. These lakes are cupped in permanently frozen ground, and exist primarily because the permafrost prevents water from filtering into surrounding soils. They are shallow, have mucky bottoms, and often are discolored from dissolved organic matter. Ponds close to rivers often receive floodwater in spring. Most are not connected with any visible stream system: often the streams meandering across one of these flats seem to miss all the lakes on purpose.

Most of the larger, deeper lakes of the region are at the edges of valleys elevated above the present river. They were formed by a combination of geological upthrusting of the land and downcutting by the river, leaving them perched behind debris dams. These lakes are clear, seasonally stratified into layers of different temperature and density, and usually have sandy or gravelly bottoms.

The delta complex where the Yukon and Kuskokwim rivers find the sea is a distinct freshwater region by itself. Between the two rivers almost half the landscape is covered with shallow thaw lakes. Many small, tortuous streams start in and wind through the alluvium of mid-delta. Large sections are flooded occasionally with silty water from mainstem rivers. Along the outer delta, Bering Sea water sometimes is driven by storms far up the coastal sloughs and into adjacent freshwater marshes and ponds.

The most spectacular family of lakes in Alaska lies on the east flank of the mountains dividing the Kuskokwim and Bristol Bay drainages. These are the Wood-Tikchik Lakes, a dozen large and several small clearwater lakes lying in deep, nearly parallel glacial troughs half within the rugged mountains. Across the pond-dotted Bristol Bay lowlands is another group of much larger lakes tucked into the Aleutian Range at the base of the Alaska Peninsula. The two biggest lakes in Alaska (Iliamna and Becharof) are among them. These and others on the west side of the Aleutian Range help settle glacial and volcanic sediments, clearing the waters of many short rivers that sing into Bristol Bay.

Because the mountains of the entire southcentral region have active glaciers, almost all of the long streams of the region are turbid and high in summer, low and clear in winter. Shorter streams from low elevation watersheds have excellent natural water quality. Their flows do not constrict as drastically as arctic and Interior streams do in winter because of periodic thaws and generally warmer air and soil temperatures.

There are over 25,000 miles of streams, but relatively few lakes, in wet and mountainous southeast Alaska. Watersheds often are steep, the soils shallow and gravelly or completely water-logged, so that stream levels tend to rise rapidly in rainy weather and drop quickly in dry spells. Peak flows of glacial streams occur in July, August and September. Nonglacial streams usually are warmer and less turbid. Their peak flows occur with rapid snow melt in April and May, and again during fall rains.

Living Systems in Fresh Water

In Alaska life in freshwaters is profoundly affected by water temperatures, light and nutrient supply. Nearly all the state's fresh waters are colder and receive less annual light and lower nutrient additions than temperate-zone waters, but local variation is great. Water temperature mainly reflects seasonal air temperatures but also is affected by shade and inflow from warm springs or glacial meltwater. Light—primarily a function of latitude and cloud cover—is affected also by ice clarity, snow cover and turbidity. Nutrient supply responds to rates and volumes of organic decomposition as well as to the content of water draining into the waterway.

Arctic Alaskan ponds and lakes have received a surprising amount of study—much more than arctic rivers and subarctic freshwater environments. All share certain characteristics: low phosphorus and nitrate supplies, which limit productivity; maximum summer water temperatures of less than 50°F; a virtual absence of photosynthesis in the period of midwinter darkness; and low biological diversity and productivity.[23]

The myriad shallow ponds of the Arctic Coastal Plain owe most of their biological activity to photosynthesis by sedges and grasses at their margins, to the organic debris from dead plants, and, especially, to ancient organic debris from thawed sediments. Algae on the bottom add moderately to pond productivity. Production by phytoplankton is low because of the very limited period when light and unfrozen water are simultaneously available, and because of grazing by zooplankton. Primary production from all sources probably is in the range of 30-50°C/m2/yr. Dead algae and sedges underpin a detritus-based food web that includes invertebrates like the superabundant midge (chironomid) larvae and the birds that feed on them. The bottom algae and phytoplankton are grazed by small zooplankton. These feed predatory zooplankton and insect larvae, with birds at the top of the food pyramid. Fish and fish-eating birds are absent (except, occasionally, as summer visitors) from ponds that freeze to the bottom.

The deep mountain lakes of the Brooks Range have extremely low primary production (1 to 8 g C/m^2/yr). They have essentially no emergent vegetation. Little debris is washed in except in periods of high runoff, but considerable organic litter blows into the lakes. Nitrate and phosphorus are at very low levels, limiting algae growth. Moss mats on some lake bottoms contribute substantial amounts of organic material. Fish like lake trout, arctic charr, whitefish and burbots are slow growing but may be common, feeding on bottom-dwelling invertebrates and on smaller fish.

Very little is known about the systems of life in Alaska's arctic streams. Larger rivers have more species of fish (about 20) than nearby lakes mainly because the rivers provide niches for species moving seasonally in and out of lakes and estuaries. River fish somehow must find suitable places to spend the winter when water is scarce and find places where food is concentrated so that growth can take place in the brief summer. Spring-fed streams are favored because they have dependable water and high insect populations. Deep pools in major rivers, where debris accumulates in summer and where water is present all winter, are heavily used. Crucial to fish, these deep pools are coveted sources of fresh water for oil operations too, and there is some anxiety about whether enough water exists for both.

In central Alaska, freshwaters share many characteristics with arctic waters: low winter light levels, prolonged ice cover, low water temperatures and low nutrient levels. They also have features peculiar to their continental, subarctic settings. Inland lake surface temperatures may rise to 75°F; much higher than in the Arctic. Although winters in the region are colder, the stored summer heat in sediments and water, and the heavier blanket of insulating snow in winter, result in shallower ice. More water is available in winter. Finally, there are more sea-run fish and emergent and aquatic plants in central Alaska's ponds.

In thaw lakes phytoplankton, rooted and floating plants, bottom-dwelling algae, and algae growing on larger plants all contribute to primary production. Nitrate and phosphorus accumulating in the lake during winter support a brief but intense period of phytoplankton growth before ice leaves in late May, and a higher peak in June when water temperatures warm. Algae production drops as nutrients are depleted. In August there is a resurgence of primary production from algae growing on the stems of sedges, rushes, cattails, pondweeds and water lilies. Those flowering plants also contribute significantly to the total primary production of the lake (a modest 20-25 g $C/m^2/yr$), most of which dies and enters detritus food webs without being grazed.[24] Fish cannot reach many isolated thaw lakes, and even if they could, they probably could not survive the low winter oxygen levels. These lakes support a moderate population of breeding and transient shorebirds and ducks.

High in the Alaska Range, the Tangle Lakes represent a very different ecological situation.[25] These alpine lakes lie in the path of several creeks that bring a continuing flow of mineral nutrients into and through them. Light penetrates deep into the clear water. Ice covers the lakes (which lie above the 3,000-foot altitude) for all except three months; still, high daily growth rates of phytoplankton in summer result in yearly production of about 100 g $C/m^2/yr$, far higher than the other water bodies just described. There are few emergent plants but many beds of submerged pondweeds. Plankton and attached algae are abundant. Grayling, lake trout, whitefish, sculpins and burbot capitalize on the populations of detritus-feeding invertebrates, large zooplankton and small fish. Diving ducks harvest pondweed during the months before freeze-up. Moose, belly deep in the lake, water dripping from each drooped ear and shaggy tuft of hair, contentedly graze the aquatic pasture on long summer evenings.

Rivers in interior Alaska have low primary productivity. In many streams bottom-clinging algae are the only important photosynthesizers. Even these are sparse in streams with high turbidity. The Chena River, which winds through Fairbanks, is typical of small, usually clear Interior rivers. It has quite abundant diatom algae stocks on river gravels. Bottom-dwelling insects are most abundant in the middle reaches of the river. Low winter oxygen levels and low total phosphate probably are controlling factors in the ecology of this stream. Fish that live in rivers like the Chena depend almost solely on detritus food webs. Attached algae, fungi and bacteria populations are high in fall and winter in many boreal streams, and, not by coincidence, fall and winter are important growth periods for aquatic insect larvae. Juvenile fish concentrate in spring-fed areas, clear tributaries and ponds where zooplankton and insects are more abundant than in run-of-the-river areas.[26]

In southcentral and interior Alaska there are hundreds of glacial streams where cold water, silt and low light reduce instream primary production to patchy growths of algae in quiet backwaters, and where few insects can survive the rush and smother of mineral particles. In those streams most forms of life seem to be transient, including in-migrating salmon bound for clear tributaries, out-migrating salmon smolts, and the fish and birds that feed on live and dead salmon. However, there is much to be learned yet about living systems in these glacial waters and their associated water-filled gravels.

Southeast Alaska's freshwaters are north-temperate rather than subarctic because of the region's milder climate. However, they often have lower productivity of plankton, large invertebrates, fish and birds than freshwaters in central Alaska. A major reason is that most southeast Alaska streams and lakes depend on recently fallen rain which has percolated rapidly through the watershed's young mountain soils. This water is very low in dissolved minerals and organic compounds: streams and lakes often are nutrient-starved. Also, the water stays cool in summer because some of it originates as snowmelt and because of the frequency of cool, cloudy days. Debris from overhanging bushes and trees, organic matter from muskeg drainage, a small contribution of primary production from bottom-dwelling algae, and a surge of nutrients from decomposing salmon constitute the basic food resources of these stream ecosystems. Because conifer needles have lower nutrient quality than deciduous tree and shrub leaves such as cottonwood, alders and willows, the kinds and abundance of life in southeast's streams can vary from headwaters to mouth, and over time as logging or other land uses change vegetative cover in the riparian zone.

When salmon bring the sea's wealth shoreward, the benefits are not only to commercial, subsistence and sport fishermen, or to leaners on bridge rails, but to entire riverine ecosystems. Predators and scavengers time their seasonal movements—even breeding schedules—to the predictable bonanza. Underwater are the egg-and-fry eaters, the sculpins and sticklebacks, charr and trout, ouzels and mergansers. Above are the wheeling, yellow-eyed gulls, the plunging terns, ospreys and eagles. And the bears: black, brown, blue, grizzled yellow. Underlying and extending beyond this raucous activity is something much more basic, the quiet but massive pulse of nutrients from dead salmon that reverberates from algae and tiny detritus-feeder to caddisfly and skimming swallow. Through river waters, the nutrients of the land gravitate seaward. Through salmon, some of them return.

It makes sense that there should be more vertebrates making a living in fresh waters that have high primary production or a high biomass of invertebrates than in less productive waters. That generalization probably is true, even if empirical evidence is very sparse. We know of a fairly common exception to the rule, which is that many shallow lakes have moderately high productivity but have so little liquid water in winter, and such low oxygen content in remaining pools, that fish and furbearers cannot live there. Another problem in trying to match vertebrate abundance neatly with productivity is that important groups like salmon and water-fowl are migratory. Their numbers at times are controlled by habitat conditions or by human harvests elsewhere. Right now, for example, continental waterfowl numbers are at their lowest ebb in decades, mostly due to droughts in midcontinent but as well to winter habitat losses, and there are few to dabble in Alaska's waters in summer.

For statewide coverage and continuity, no other studies match the aerial surveys of waterfowl breeding populations done for the past three decades by the U.S. Fish and Wildlife Service. Even though questions remain about the accuracy of the estimates their precision and hence comparability from area to area are good. They clearly illuminate regional differences in waterfowl densities, ranging from about 10 to 100 ducks per square mile.[27] The two areas with sparsest duck populations (Cook Inlet, including the northern Kenai Peninsula and Susitna lowlands, and the Arctic Slope) and the two with the densest stocks (Yukon Flats and Copper River Delta) illustrate some of the important factors in duck distribution. Duck habitat on the Kenai Peninsula consists mostly of medium sized, cool and fairly deep lakes with few marshy shallows. Neither the size and depth nor the productivity levels of these habitats is favorable for tightly spaced waterfowl breeding territories. The Arctic Slope, in contrast, has good habitat geometry (numerous shallow ponds, closely spaced), but relatively low productivity of edible plants and invertebrates. The Yukon Flats and Copper River Delta are very favorable in both respects.

Phosphate and nitrogen compound levels in lakes and ponds parallel the abundance and variety of summering ducks, at least in the Interior where recent studies were done. These chemical levels are linked to the presence or absence of direct water connections between the duck lakes and nearby creek systems. Lakes connected to creeks have higher nutrient levels, high plankton populations and more food for ducks than isolated lakes.[28] (Of course, other forces may be at work, such as varying predation rates or different exposure to flooding; for now the nutrient-supply relationship seems the best explanation.)

Scientists have not yet developed much information on produc-tivity of all fish living in a particular arctic or subarctic environ-ment, but studies of a few individual species give us an idea of what to expect from the fish fauna as a whole. Burbot, for example, which eat invertebrates until switching to a fish-based diet at about three years of age, grow more slowly than the same species in temperate North America. They reach sexual maturity later, too: 6-7 years instead of 2-4 years. However, Alaskan burbot may live longer and eventually get bigger than their southern counterparts.[29] Similarly, lake trout (lake charr) generally grow slower, mature later and live longer than lake trout in southern Canada and the northern tier of U.S. states. Moreover, within Alaska the same differences occur in stocks in different environments. In the downstream, warmer of two large lakes in the same drainage in southcentral Alaska, lake trout mature slightly earlier (8-9 years) and live a year or two less long (almost 15 years) than their upstream relatives.[30] In the cold waters of Brooks Range lakes, the lake trout live a good deal longer. In one sample from Chandler Lake, 58 percent were over 15 years old, 32 percent were over 20, and one was 42 years old. These far-north lake trout, once mature, usually breed every other year.

Among grayling this pattern is repeated. Grayling in the Bristol Bay drainage are the growth champions in both ultimate size and yearly gain. Not only are their home waters warmer, but they are also host to enormous salmon runs which provide food to grayling directly (salmon eggs) and indirectly (aquatic invertebrates in the salmon-carcass food chain). In the Brooks Range grayling are by comparison very small, slow to grow, and long of life. Grayling of the Interior are intermediate.

Dolly Varden charr reflect these same environmental differences. One student was able to show that even among charr from the same stock in the same stream, individuals that lived in warm and productive beaver ponds grew far bigger than siblings in the stream a few meters away.[31] As some coastal Dolly Varden are resident in freshwater while other stocks, the majority, are anadromous, it is possible to see differences between fresh and saltwater environments in the same region. Charr that spend months feeding in saltwater grow faster and bigger than landlocked charr.[32] However, all overwinter in fresh water, a habitat that may be safer from predators.

Subarctic and arctic fresh waters, limited by highly seasonal light, by low levels of nitrate and phosphorus, and by cold, simply do not match the productivity of their temperate counterparts. If you compare similar types of rivers, streams, ponds and lakes in different regions in Alaska, production is higher in more southerly areas. (Except that southeast Alaska's fresh waters are often less productive because they are more nutrient-limited than those of the Interior.) The amount of animal life supported by the year's photosynthesis within the lake or stream is often much less than the amount supported by new or ancient detritus that blows, falls or washes into the stream. Alaskan rivers, in particular, are usually quite unproductive, scoured as they often are by severe floods, starved for water in winter, and hissing with suspended silt. They do serve, however, as highways for moving fish and as havens for overwintering stream fish.

3

Messages from Earth

I love to try to read landscapes. As far back as I can recall, I have been intensely stimulated by the sight of (to me) new wild and rural places. Curiosity gets into gear first, and dozens of questions race through my mind. I wonder how far back these side valleys go? Does that stream ever flood—yes! Look at those ice scars on the willows and cottonwoods! What kind of flowers are those? I'll bet there are deer in that old burn…and on and on for as long as I'm able to be alert to the country. Almost as soon as my interest is engaged, it becomes mixed with a more personal involvement. I ask "What would it be like to live here? Where would I want a home, by the creek's music or up where there's a view? Would a driveway be too hard to build and too erosion-prone around the nose of that hill?" Like an old house cat trying one pillow and pile of clothes after another, I start nestling into the country, trying it on for comfort. (About this time my wife is asking whether I want her to drive.)

Similarly, though in a much paler, more abstract way, our brief look at the nature of nature in Alaska is at once simple descriptive geography and a statement of potential meaning to people. Essentially every descriptive comment is a message from earth to us, whether I have been able to read it or not: make sure we and the organizations tending our relation with nature are flexible enough to survive in diverse and changing environments. Reckon on century-long replacement times when planning timber harvests. Design oil rigs to withstand extreme wave forces in the Gulf of Alaska—or don't put them there at all. Don't expect to graze as many animals on wild tundra ranges as on temperate grasslands. Ration an angler's catches of trophy trout—they take a score of years to replace. Invest sparingly in structures on earthquake faults. These and hundreds of similar messages can be read as guidelines toward confident and sustainable northern living.

In this chapter I'll lift out, reorganize, and comment more pointedly about the messages I'm able to hear. Later (Chapter 5), I'll take a step toward guidelines for action, the kinds of human behavior and social strategies that seem consistent with these messages. For now I'll center on the elements of the natural character of the North that are so rich in implied human meaning.

Every Place Is Different

Diversity is a slippery subject. It remains to be seen after I tackle its slopes whether I end up on my feet or tail. It can't be ignored, however; *difference* is too universal and important.

With regard to diversity, there are splitters and lumpers. Splitters follow through layer after layer of detail from statistically significant differences in proportions of plants on neighboring slopes to facts of microscopic anatomy and find—to no one's surprise—that every place and thing is different. There is so much variety in nature that the only way a place could be unique is for it to be exactly like somewhere else. And then there are lumpers, who move upward into the thinnest atmospheres of generality, and conclude that all places are the same because they consist of matter and energy moving around in similar circles. "When you've seen one redwood, you've seen them all."

My interest here is in those levels of geographic diversity that are of practical importance to people. One place is different from another if we can't behave the same there as in another place without imposing extra costs on ourselves or the rest of nature in the long run. Because there are different levels at which we can describe and control our behavior, there are different scales at which we define places as different. If we want a forested landscape to remain forested indefinitely even while we take trees annually from it, we can start with the broad guideline of allowable cut: the wood taken out can't exceed what is added annually by growth. At that level, all harvested forests are alike. But to make that work for a particular species of tree requires responsiveness to its characteristics.

One species can be perpetuated under a clear-cut harvest system, another only under individual tree selection or some other technique. At this level we might be able to say that all redwoods truly are alike. But to translate this to a viable long-term prescription for forest management demands a focus at least down to the level of a stand, a group of trees growing together in a place of similar soil, weather, exposure to sun and so on. The scale has tumbled in this illustration from "all forests" to "all redwood forests" to "this stand of redwoods," each level expressing a kind of diversity that demands response.

Desired scale helps define the limits of homogeneity, and so does the characteristic whose variation is important to you. A farmer may be most interested in soil characteristics and the seasonal incidence of frost. What wild animals live in a place or what wild plants now grow there are secondary, functioning as indicators of soil condition. A builder of roads and pipelines, likewise, might want to know how permafrost is distributed, not what precise plant community is there. People use criteria for environmental diversity consisting of scales of flood frequency and severity, risk of ground movement during earthquakes, presence of potable or dammable water, and so on. In short, when you speak of places being different, you need to know what the test of difference is. In the first part of the book, my implicit scale was an unspecified combination of natural characteristics (sea ice presence, climatic regime, fire frequency, existing vegetative cover, and primary productivity, to name a sample). Most attempts by ecologists to define natural regions and subregions—and mine is a very loose and subjective one—focus on some combination of these; a geologist would use almost totally different criteria.

And so, at various spatial scales and using a loose collection of environmental criteria, Alaska is full of places different from each other. In a word, it is diverse. Why does it matter?

Most people look at the world accessible to them as a collection of human possibilities. Diversity increases the number of possibilities. From one ocean environmental complex comes the possibility of pollock, from another salmon, from still others petroleum, gold, whales and tidal power. Northern landscapes offer different combinations of values of copper, caribou, the energy of wind, sun and river, the scent of wildflowers, physical challenge, adventure, home sites, oil. The total diversity of what human politics has defined as Alaska is immense, a product of both size and basic factors such as atmospheric and geologic variety. On whatever scale you focus, diversity reduces the need for imports, increases the chance for exports, and buffers human communities against economic extremes. Because these environmental values take different skills to use and different personalities to appreciate, and because different lifeways grow out of the interactions of particular resources and their users and tenders, human society itself becomes richer in variety and opportunity.

Differences in environmental and resource character lead naturally to the need for differing, problem-tailored stewardship strategies. Suppose, for example, that of two bays in the same sound, one is sheltered and comparatively shallow, receiving the flow of a river winding through a delta marsh into the salt waters. The other is exposed to sea swells, has steep rocky edges, and is turbid with silt from glaciers at its head. Both have great human value even if the sources of value differ. At this stage in the development of our environmental caretaking strategies, we would try equally hard to prevent passing tankers from spilling oil into either bay. If the beaches of both bays were oiled, however, we would spend more effort cleaning up the biologically richer sheltered bay simply because, given limited cleanup resources, we would try to gain the most benefit for every dollar spent.

We make these and similar choices constantly, trying to match stewardship technique and intensity to the varying natures of the risk and the land. When the Trans-Alaska Pipeline was being built, the people responsible for fish and wildlife protection gave priority to stream crossings, big game crossings, and slopes and soils very vulnerable to erosion and situated where eroded soil could reach streams. The number of people devising and carrying out prescriptions for solving stream crossing problems varied with the specific landscapes and their perceived values to us. Much the same pattern of thinking has resulted in regulatory strategies reflecting the responses of various soil and vegetation types to an array of off-road vehicles, and differences between snow-covered and snow-free surfaces. In yet another example from neighboring Yukon, streams are classified as to their biological richness and potential vulnerability to increased silt loads. Their class determines whether placer mining is prohibited or permitted with differing degrees of restriction.

Just as caring for differing landscapes requires particular rather than general knowledge, so does building in them. In core design or in small detail, everything we build needs to be tailored to the character of particular sites. Along a well-travelled road in suburban Fairbanks, there is an ordinary house that is now a tourist attraction. Nothing about its architecture suggests why tour buses should stop beside it, and its grounds are unremarkable. No one famous ever lived there. What titillates the interest is the combination of tipsy angles made where the main house meets an offset el, where the porch meets the house, and where the attached garage joins the main structure. It is—the tour guide explains—a classic case where someone built unknowingly or uncaringly on frozen ground containing blocks of ice. Bulldozed clear of its plant cover, the site warmed in the sun, the ground thawed and slumped, and inevitably, the foundation of the house followed suit. Now the house stands drunkenly on collapsing soil, empty and unsalable. Unofficial estimates suggest that there are about 100 other homes in Fairbanks with similarly fatal problems.

Diversity also matters to us as persons. Humans have wonderfully wide-ranging physical and mental capacities that, to be discovered and expressed, need the stimulation of diverse environments. In a varied environment one can follow deep-held preferences toward places whose spirit matches one's own: close-cupped glades, the dynamic border between green land and blue sea, the free perspective of hills. Alaska certainly has that variety. I remember an early August day in Juneau when I climbed to a high glacier edge one morning and descended to the waterfront by evening to cast for salmon, with long intervals of forest traverse between. It was a marvelous day full of surprises, interest and small adventures. Diversity presents physical tests, exercises in observation, countless small decisions of adjustment to barriers and beckonings.

As individuals and in our corporate and government organizations, our record of sensitivity to environmental diversity is mixed. Overall, I think there has been a lot of improvement in the 30 years of my residence in Alaska. To the extent that laws and bureaucracies have been created by people strange to the North and driven to express needs "on average" over the whole nation or the whole state, we have been harassed by blunders in trying to live and earn livelihoods in distinctly non-average places. In the fall of 1989, a report by the Inspector General concluded that $300 million had been spent by the federal government on just under 3300 homes for rural Natives in Alaska, a staggering proportion of which were either ill designed, poorly built or badly maintained—and falling apart. Granting that many other factors contributed to the failure of this program (or even allowing that the program might not have been a failure overall, but merely inefficient), it is clear that unfamiliarity with a northern climate and landscape on the part of standard-setters, designers and construction bosses led to a fundamental mismatch of intention and execution.[33]

Conversely, to the extent that national or statewide policies have reflected northern realities or given regional officials latitude to tailor programs to those conditions, and to the extent that people familiar with the North have carried out the programs, the fit has been good.

Individuals—some individuals—have tended to respond more quickly to environmental diversity for the simple reasons that their focus is local, and they don't always have to operate from a consensus position of big groups of people. Some local fishermen build boats to suit the water they fish in. Gardeners choose the sunniest places on their property. Home builders avoid obvious hazards like avalanche tracks and places in the reach of storm tides. Obviously, individuals continue to make mistakes, too, out of ignorance or willfulness or lack of choice. Ignorance, I think, and perhaps willfulness (or at least narrowness of purpose), have been made worse by the high proportion of newcomers and transients in

Alaska's society this past century. Many people haven't had time or strong incentive to learn the country.

Places Don't Stay the Same

Change across distance is matched by change through time, in Alaska and everywhere else. Temporal diversity, like diversity of place, is inherently a matter of scale and intensity: slight changes that accumulate slowly over big areas, catastrophic changes that are very local, and so on. From the standpoint of understanding nature and exerting control for human benefit, it pays to try to categorize changes in their relation to the interplay of life and the physical earth that is all around us. For instance, there are changes that are so recurring, frequent and cyclic that they almost fall out of the idea of change: tides, day and night, seasons and common weather patterns are examples. Transplanted (exotic) animals or plants might have trouble coping—there are temperate-adapted vegetables that survive in Alaska but bolt or fail to set fruit because of the length of summer daylight; and there are non-Native people who suffer medically treatable mental depression in our long dark winter months—but indigenous people don't.

Then there are changes that are less frequent or regular, but don't cause major system upsets. Forest fires, the rare warm spell in winter that fools some plants into sap flow, and floods are examples. Fires and floods are fatal to some individuals and may measurably disrupt everyday system function, but recovery is predictable and draws from locally available (perhaps neighboring untouched) parts of the system itself. These events often act as evolutionary selective forces molding system character.

Finally, there are changes we can call catastrophic or cataclysmic in that they result in brand new kinds of ecosystems. They range from earthquake-induced salt-marsh subsidence that turns a wetland community more or less permanently into a subtidal one (a rapid phase-change) to the cumulative effects of climatic change in which over several thousand years a forest-dominated system gives way to grassland. What was once there may never be restored. However, as long as a place is habitable, it will be inhabited. The question is, by what?

These very questions are deeply involved in the turmoil among resource agencies and interest groups following the massive crude oil spill from the *Exxon Valdez*. We don't know whether intertidal and ocean-floor biologic *systems* have been affected or whether the disruption is best described merely as varying intensities of damage to particular animals. The first view implies a systemic, deep, structural disruption and clearly assumes that wholes exist that are more than an accumulation of parts. The latter view is more atomistic, centering on quantitative changes in stocks and downplaying interdependencies; it denies or shifts attention from higher levels of biotic organization. We can't accurately measure damage or even define it and have little information about what was there before the spill. We don't know whether individual otters, limpets, clams, etc. will recover or how long before they are replaced by a new generation. And we don't know whether the kind of beach community that is reestablished in a particular cove will be the same as the one the oil smothered or poisoned. Probably more thoroughly than any other recent incident of Alaskan life, the *Exxon Valdez* spill abruptly tore the concepts of resilience, diversity, stability and recovery out of the ivory towers of academe and set them thudding on the desks of lawyers and society's pragmatic leaders.

These differences in scale and intensity of change over time obviously have consequences for management strategies. For example, it is probably sensible to assume that a forest working circle (a bureaucratically defined area within which allowable tree harvests are calculated to match sustained yield targets over time) in southeast Alaska will yield a calculable and dependable amount of wood fiber for harvest yearly over a century-long rotation; both the planning horizon (100 years) and target yield are biologically acceptable. In the Interior, unless fire suppression is near-perfect, the planning period is too long. And in a typically fluctuating oceanic fish stock, neither the 100-year framework nor the fixed average harvest make any sense at all.

Biological Wealth Moves and Concentrates

Because places vary in temperature, water regime, nutrient supply and many other factors, and because of huge seasonal and longer term fluxes in environmental quality of a particular place, life flourishes in the good places and times, and avoids, leaves or does poorly in the bad. I've given some examples already. Let's look a bit further and see what they mean.

So often when people encounter wildlife, the animals are on the move: the sow grizzly browsing, shuffling or galloping its way to the far reaches of its enormous home range; the skein of geese arriving from Sacramento rice fields; the grouse that hits the windowpane in late September; the silvery salmon dipped from a swift and roily river; the woodfrog crossing your lawn far from its spawning pond. The scale of these movements varies from intercontinental and interoceanic to local. Reasons for moving are equally varied.

Caribou of the Porcupine caribou herd of northeast Alaska and Yukon nicely illustrate some of the diverse but important reasons for wildlife movements. These ungulates lick their newborn calves clean in June near the Arctic Ocean in the Arctic National Wildlife Range, gallop to the breezy coastal points and islets to escape insects in July, joust for mates on the flanks of the Brooks Range in October, and bury frosty noses in the shallow snow of Yukon's boreal forest two months and two hundred miles later. In the broadest of terms, they move primarily to reach calving grounds where predators are scarce, to find the most nutritious and available food at each season, to reduce insect harassment and to avoid snow conditions that would hamper their escape from wolves or use up precious energy stores too fast.

Stories could be told about scores of other arctic and subarctic animals in their journeys of survival. Among my favorites would be the curiously complex movements of Dolly Varden charr and certain arctic whitefish between lake and river and ocean, the swift wave-stitching flights of shearwaters and petrels across the North Pacific, and the clamorous tide of sandhill cranes that flows into Alaska every May and ebbs again in September. As more are told, the tales begin to weave broad patterns: patterns describing limits of animal adaptation, life's infinite tricks of survival, and interdependence.

What meaning does this have for us? The crucial message, I think, is interconnectedness. Through a chain of ecological transformations, upwelling phosphorus in a storm track near the Aleutians is taken into—becomes—the muscles of a salmon that swims to the slender beginnings of the Yukon River. In the same near-miraculous way, the dry arctic grasses grow hooves and walk, and the flesh of tiny ghost crabs on beaches of the central Pacific lift new avian wings each spring toward the cool tundra marshes of Alaska. Thus, and in countless additional ways, Alaska loses its illusory isolation and becomes what it truly is, a hub in the living traffic of the whole Earth. As we try to use and care for this mobile life, we, too, become intimately interconnected to people and life beyond our everyday boundaries.

When in transit, migrating fish, birds and mammals often move in big groups or collect together in favored places en route. In such places they need safe haven and good and ample food. Whether they find what they need is likely to determine how well they cope with the rigors of the next life cycle phase, whether that is the frenetic breeding season or the long winter maintenance interval. Such a place, for sandpipers, is the Copper River Delta. There, in spring, the deposited silts from far inland yield their bounty of burrowing invertebrates to between 10 and 20 million dunlins, western sandpipers and their kin. The birds have come far, and their fuel supply—fat—needs replenishing if they are to complete the last lap to their still frozen nesting grounds and bring to the intense breeding-season activities the vigor essential to success. Thus they, like other animals on the move, travel spaces described not by all-inclusive outer boundaries but by special travel-ways, funnels, gathering places, cunning synchronies and crucial timings. And in our relations with animals on the move and in their gathering places, we have to remember that all the mothers, if not all the eggs, may for a time be in one basket. Mostly we think that it is scarce animals that face extinction, but even abundant ones may place their existence at risk in one time and place when, biologically, they have no choice.

A person wanting to harvest plants or animals that move around and concentrate needs to know where they are, to be able to live or get there for harvests, and to have the opportunity to do the harvesting (which, variously, is a matter of custom, land ownership or legal process). Mostly we know where and when biological wealth builds up, in these days of remote sensing, seasonal resource inventories and intensive, rapid communication. Preindustrial people lived close to where animals were abundant;

sometimes this meant moving seasonally or every few years. Now it is technically possible to intercept wildlife almost anywhere in Alaska from any home community (even as far away as Germany, the source of many hunters of big game in Alaska). The advantages of long distance access are obvious, but there are serious disadvantages too. Any place you can reach, so can hundreds of strangers. And the main criterion for modern access is wealth: ability to buy airplane tickets, jetboats, snowmobiles and so on.

For those who try to manage wild things and the people who use them, the pervasive habit life has of flourishing or appearing at certain times and places means that research and regulation have to be specifically focused. Usually the critical problem for the manager is the ability to control human and sometimes natural forces at high-value sites. To reach their targeted control, agencies have many arrows they can try, from use permit stipulations to legislation and from public education to interagency agreements. Most of the arrows, unhappily, are warped or blunted.

Earlier I mentioned the nomadic caribou of northeastern Alaska, and it happens that they present typical stewardship problems of animal mobility and concentration.[34] The outer bounds of the area these caribou have visited and used in recent times enclose about 96,000 square miles (Figure 5). The herd usually winters south of the Brooks Range in the south and southeast sectors of their home landscape. In spring and summer they are in the north, often close to the shores of the Beaufort Sea. As spring approaches the cows and young animals trot ever more determinedly toward the north flanks of the mountains, where, somewhere within a fairly extensive section of foothills, they will collect together quite densely in a chosen area. The pregnant cows will give birth and begin the critical process of nibbling selected foods to rebuild their strength. Exactly where they will calve is a matter of guesswork for humans and perhaps of last-minute choice by the caribou. Probably the vagaries of late winter snow distribution are important not only in the site that is optimal, that year, for calving, but along the whole migration path. The result is that when you map the places selected over a number of years, you see that some are used quite often and others occasionally. Any of them—and others—could be selected in future years.

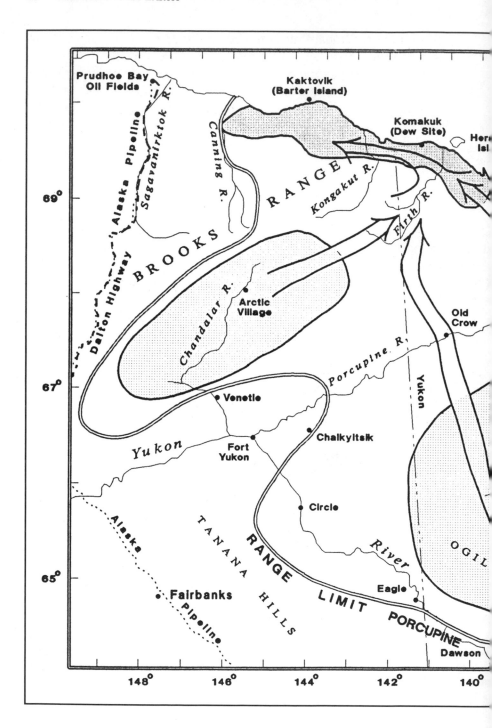

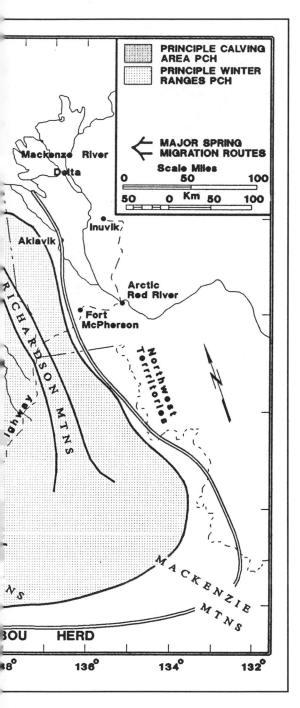

Figure 5. The range of the Porcupine caribou herd in Alaska and Yukon. Courtesy of the Alaska Department of Fish and Game and the United States Fish and Wildlife Service.

In the debate about exploring the coastal plain of the Arctic National Wildlife Refuge for oil and natural gas, all sides at least outwardly agree that caribou calving grounds need special protection. Some would be satisfied with a short, calving-season restriction of aircraft and ground activity in the prime (often-used) areas. Others think in terms of keeping the calving grounds off-limits all year to petroleum-related activities. Either way it is clear that over the life of an oil field (say, 30-40 years), the calving grounds are moving management targets. A broad-brush withdrawal of the whole coastal foothill and plain province of the Refuge would assure protection to the caribou every year but prevent oil exploration. A year-by-year floating sanctuary would give oil interests greatest freedom, but the intensity of protection caribou would get, as well as *area* of protection, would be far less because fixed structures like roads, pads, airfields, wells, buildings and pipes wouldn't be packed, moved or shut down when, some spring, the caribou showed up there.

The Opportunists

Another characteristic of northern plants and animals is that quite a few are opportunists. I've seen many examples. Think of ravens whose days, it seems, are a far-ranging search for anything useful: a moose killed and only partly eaten by wolves, an overflowing dumpster, a bag of groceries dropped in a parking lot, or a plume of warm air from a power plant in midwinter, where the sable birds soar for long minutes at a time. Again, consider migrant bank swallows that dig their tunnels in steep silt walls formed only a few weeks or months ago by river cutting or road building. In a young landscape where geologic forces are strong, in forest-fire country where plant communities are snuffed out, regrow and are burned again, and in a land where basic prey species like lemmings and hares alternately teem and disappear, opportunism is an obvious strategy of repopulation and survival.

Two different kinds of opportunism are illustrated by fireweed and the pastel-colored corydalis, both common in the boreal forest zone. Fireweed is well named. In the intervals between wildfires, its tall stalks bearing pink-red flowers are common along roads and riverbanks where the soil has been exposed, and it also is widely scattered through woodlands, often very spindly and weak in these shaded environments. Let a fire burn the woods, however, and fireweed is suddenly everywhere, often forming an almost pure stand of flowers a year or two afterward. These plants start from two sources. The surprisingly thick and fleshy roots of fireweed lie in the soil just beyond the depth that fire heat normally reaches, and they sprout vigorously a month or two after a spring fire or the spring following a fire in July or August. Many of the colonists have travelled a long way, however. There is a time in late August in the Interior when snowstorms of fluffy fireweed seeds drift across the fields and sift through the forests. Landing on ash-covered ground or bare earth, they are ready to take root the following year. The northland's best-known flower, then, is both an endurer and an invader.

In contrast, corydalis is the endurer's archetype. Consider this story:[35] On May 28, 1983 the broad sweep of mature spruce and birch forest along the highway between Fairbanks and Nenana (and, ultimately, Anchorage) had stirred itself after a long and fairly dry winter. A warm breeze blew all day, and ground vegetation, trees and air were dry when, the next day, a man fired his land-clearing debris near the mouth of Rosie Creek close to the flood-plain of the Tanana River. The fire escaped and headed west, and in spite of quick response by state and federal fire suppression crews, it burned for over two weeks, covering almost 8,600 acres. Within days the seeds of golden corydalis, which had lain dormant since the last fire in that forest at least 160 years before, began to germinate. By August these flowers were thick in the burn. (By then, too, some avian opportunists, mostly black-backed and three-toed woodpeckers, had flown in to harvest bark beetles, themselves famous followers of woods-fire smoke.)

The white-winged crossbill is another avian opportunist. Though it can and does eat quite a variety of plant foods, its specialty is the seeds of conifers which it neatly plucks with its tongue while its scissored bill lifts the covering scale. In Alaska spruces are its most abundant seed source. The trouble is that the spruce in a given area tend to have single years of high cone production (cones mature in their second year) followed by at least one year of no or low production. Trees experiencing the same weather patterns tend to have synchronized production, which means that over the Interior as a whole, there will be districts of abundant cone crops and, in the same year, areas of poor crops. The crossbills survive by wandering. They roam over large areas in small to medium-sized flocks looking for spruce with abundant mature cones. They seem to be able to begin nests at almost any season—the bird hasn't been studied closely, and there is a lot of guesswork involved in natural history accounts—and hence are able to nest shortly after discovering a good food supply.

Willows are superb models of resilience. There are, to begin with, at least 33 species in Alaska and many debatable varieties, many of which hybridize. In something as basic as genetic variability, willows show themselves to be marvelously sensitive and responsive to environmental diversity. Many species are favored foods of northern animals: willows are critical in the food economy of moose, hares and ptarmigan and are seasonally important to caribou and muskoxen. One reason is that buds, leaves and twigs in many willows are highly nutritious and are not guarded by unpalatable chemicals against the appetites of mammals and birds the way birches are, for example. Instead, willows stripped of leaves, decapitated by moose and bud-nipping ptarmigan, or girdled by hares respond by vigorous regrowth of new sprouts and twigs. When willows along a river are undermined and sent tumbling downstream, they are often stranded on bars and partly covered with silt or gravel—and promptly sprout again to take up life at a new location. And finally, willow seeds are light, fluffy and borne on even small breezes to either a feckless end on a dry upland or a new birth on moist and freshly exposed soil.

Plants whose seeds can wait indefinitely for the rare moment of suitable germination environment; plants whose seeds ride the far winds; birds that wing over leagues of uninhabitable territory in search of the time and place when food is on the table: these are only a few of the opportunistic strategies of northern life in a country where very little is ever stable and certain.

Low Rates of Annual Interest

The productive capacities of relatively undisturbed northern ecosystems were discussed in earlier chapters: the general principles are worth reiterating. South of ice-dominated and rather impoverished arctic waters Alaska's seas are quite productive, but our fresh waters have quite low annual production and would be even less productive in some regions without the seasonal runs of sea-nurtured salmon. Terrestrial systems, too, are on the slow track of energy conversion except briefly after a wildfire. In other words, the annual interest on the adult biological stock is quite small.

The stock itself—the biomass of living things at a particular moment—can be misleadingly big. The mass of soil invertebrates or crustaceans in pond water, or lake trout stocks or standing volumes of Sitka spruce, may compare with those tallied in more temperate places. The difference is in growth and reproduction rates. Animals and plants in the North often are long-lived but slow to grow or have offspring.

There isn't necessarily a direct relation between overall natural productivity and production of animals or plants valued in human economics. Not only how much, but what, is produced must be considered. For example, a tract of interior Alaska muskeg and another of North Slope tundra may both have similar rates of carbon fixation (perhaps 40-80 g $C/m^2/yr$). In the muskeg, however, the energy and nutrients are channelled into black spruce and moss, neither of which is very valuable to wild grazers, livestock or people. In moist tundra incoming solar energy is temporarily trapped by sedges, grasses, forbs and lichens that can be converted to harvestable protein by caribou or reindeer. A roughly comparable situation apparently occurs in the Bering Sea. In the southeast segment the organic fallout that supplies food to bottom dwellers is largely used by molluscs, worms and crustaceans, a high proportion of which support predator webs of bottomfish, mammals and people. In the North, however, close to Bering Strait, many of the molluscs are eaten by sea stars that, being so full of hard calcareous parts, are not often preyed upon. Except for the spawn they release into the currents, sea stars are almost dead ends in the food web.[36] In this same region the water is normally so cold that flatfish—which keep much of the bivalve-trapped energy in circulation farther south—rarely figure much in food webs.

It seems deeply ingrained in Western culture that we chafe against any sort of natural limits, and most Alaskans aware of their homeland's low biologic productivity are eager to do something about it. That "something," in general terms, can take either of two forms: adding nutrients or heat to relax natural shortages, and rechannelling energy from less useful to more useful forms. Northern farmers do both by clearing natural plant cover (which warms soil), adding fertilizers and planting a crop to capture the sunlight and soil nutrients. The Forest Service in southeast Alaska has long tried to harvest ancient mixed growths of hemlock and Sitka spruce in a way that favors reforestation by the more valued spruce. Alaskan sport fish managers routinely have poisoned existing mixed-species natural stocks of fish containing only a small portion of wanted species, restocking later with trout, steelhead and land-locked salmon that anglers cherish. Mariculture is another example. There are risks, of course; the stocked fish may not reproduce and hence require annual replenishment from hatcheries, the Sitka spruce might fall out of economic favor before its harvest age is reached. Nevertheless, these manipulations of nature have a long human history. The North undoubtedly will see its share. Taught from birth the myth of infinitely expanding economies, we shy from the obvious and much cheaper alternative: reduce our expectations of nature.

Part Two

The Learning Process

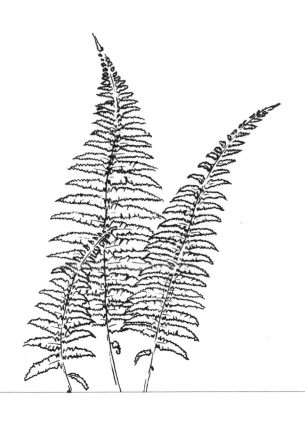

The sea and earth hum with messages of permission and constraint, invitation and warning. Are we listening? Are we learning? We transplanted southerners have come into the northern country for over two centuries now, advancing as a tide into a bay, each wave throwing more of itself forward than settles back. And spasmodically through that time we have begun fitting into the country, more quickly and quietly as individuals, more slowly as organizations and governments. Gardeners in Fairbanks now plant north-adapted varieties of flowers and vegetables. New homes, mostly still mimicking temperate suburbia, at least are better insulated. Residential lots cost less if underlaid by ice lenses.

Yet mistakes we know—having made them repeatedly—as well as mistakes we still don't admit, are legion. Nearly every daily newspaper alerts us to them. In late June 1990, for example, as I edited this chapter, the daily Fairbanks paper carried an article about a state legislator who wanted state help to move or protect a subdivision from the free-swinging Matanuska River. The subdivision, built with borough approval in 1979 alongside a river channel, had suddenly awakened to the threat of the Matanuska's penchant for soft silt riverbanks. Since 1980, the article said, the river had swallowed a swath of land some 180 feet wide and half a mile long. The governor has demurred, saying flatly that the history of the river's meanderings is well known and that the borough never should have permitted homes to be built there.

A full assessment of our actions would be enormously complex and lengthy. All I can do here is to survey briefly how we extract and harvest natural resources, assuming that it is in those processes that, having the most obvious need to adapt to local conditions, we show our knowledge and will most clearly. How well are we fitting into the country today? I know of no good standards for judgment. What is the fitness test? Is it the maintenance of natural diversity and primary production? If so, on what geographic scale? Is it the conservation of soil and scarce soil nutrients? The long-term and true monetary cost of producing a unit of timber, fish, copper, or oil? The maintenance of high populations of popular and useful animals like moose and salmon? Are there esthetic standards for northern landscapes? Do federal water and air quality standards bear any relation to what is right and humanly beneficial in the northern context?

Such standards can never be designed in a single purposive project or act of will, nor adopted by a single vote of the people. They will emerge fitfully and in fragments out of struggles to solve particular questions, and they will be adopted in equally small bits in law and custom over a long time span. That process has begun.

The best I can do here is to apply a mixture of personal bias, common sense and the best guesses of scientists to the quality of our interactions with the countryside. Perhaps arbitrarily, I assert that when gold mining severely reduces the number and kinds of animals that can live downstream, or forces us to await a century-long healing process before birds sing again in streamside willow thickets, then our mining is not congruent with nature. If our best government-controlled timber harvesting results in the total and permanent disappearance of old growth ecosystems from managed areas, then I assume that our government's management behavior needs changing. If it takes a million-dollar infusion of public subsidies and loans to get one grain farm in operation, I cannot believe our agricultural policies are optimal for the long haul. And I confess that whenever I see a sporty three-wheeler spin broken flowers behind as its driver forces it up a steep tundra meadow, I'm ready to say that we are not acting as we should toward the land that nurtures us.

Judging can too easily become blaming. Although some would argue that blaming is important because progress rides on the back of recent guilt, I prefer a more positive approach. In the second chapter of this section, I attempt to draw from the examples given and from my general experience some operating principles for nestling more comfortably—and possibly for surviving longer—in the North. They are distillations by one person from part of the range of trial-and-error experiences of Alaskans in the past few decades. They are guidelines, not dogma.

4

Gold, Grain, and Slow-grown Wood

A great many people—insiders and outsiders—think of Alaska as a storehouse of crude oil, unrefined minerals, standing timber and catchable fish, waiting to be unlocked for human use. On the cover of Joel Garreau's "The Nine Nations of North America" Alaska is mostly in the Empty Quarter, whose map symbol is a power shovel.[37] The image of the northern cornucopia has lured Euro-Americans to Alaska for more than a century. Even today every politician in every Alaskan election insists that "Alaska's future lies in resource development."

It seems fair to ask how we think of ourselves in relation to those resources. Do we see individual resources, or constellations of interacting resources? Do we invest in knowledge and steward-ship? Are fish and trees and mountains of rock totally defined by utility, or are they things in their own right? Are they valued gifts, or material things of no worth until our labor and skill transforms them?

I cannot answer those questions, I can only give impressions. Out of many possibilities I have chosen to talk about our commer-cial use of trees, about placer mining, and about northern agricul-ture as examples of our present relationship to nature. Together, these provide a chance to look at annual crops in a cultivated environment, once-a-century crops in a wild setting, and the with-drawal of a finite, exhaustible resource.

Eagle Creek: Two Visits

June 1956: In the peculiar submerged quiet of a windless mid-night, the sky luminous from the sun now slanted below the hills to the north, I climb up the small sharp hillock of stones at the edge of the creek. Gravel rattles briefly under my boots, dying away as I sit on a thin grassy tussock cowlicking the conical rise. A train of mosquitoes, hurrying to catch up, sets up a small hum around my head. Their sounds swell and wane as my attention shifts outward and inward. So, too, does the washing of the creek, sometimes smothered as if by a blanket of heavy air.

Most of the time I watch the sloping shoulder of a nearby hill of gravel, upstream. There, touching with its slatey feathers an undis-tinguished stone of gray schist, a wandering tattler broods its blotched greenish eggs. Three weeks earlier it had completed its northward flight from some bright Pacific island beach, with a way-station for refueling on a rocky Alaskan shore. Soon its four blue-gray chicks will break the walls of their oval world, toddle precariously the two yards to streamside and nip curiously at the wet-skinned insect nymphs that will fuel their growth all summer.

Upstream and down the humps of gravel—tailing piles of mines abandoned before I was born—define the channel of the small clear creek. A few of the hillocks, made of coarse gravel and rocks, are bare except for young lichens clinging to the stones. Most have a scraggly mantle of willows and alders, patched with lupines, dwarf fireweed, lousewort, moss campion, milk vetch and other alpine flowers. Thickets of willows look oddly blunted and broken where moose had stopped to browse. I know that if I look at other willow branches, I will see where ptarmigan had nipped leader buds early in May, forcing a shoot to form from a smaller side bud. In a few minutes a gray-cheeked thrush flies like a quick shadow into an alder thicket. A small grayling flicks a spray of water into the evening as it dives with a stonefly in its mouth. The slim valley bottom is bordered on the south by a brush-bordered ditch that once carried water to a mining penstock down-valley, and on the north by a wagon trail. The knee-high tundra wildness begins just beyond.

October 1985: All summer since the creek's winter ice broke in May, and every past summer for over a decade, the bulldozers had filled the hills with their roaring, creaking flatulence. This gray October day, though, they are quiet. Freeze-up has locked away the water; gold cannot be separated from the gravel and the miners have no more business with the creek. Upstream from where I stand, hands thrust into my down jacket pockets, the creekbed runs like a dull brown gash into the flanking dun hills. Downstream it is the same. Piles and streaks of unsorted gravel, mud, roots, peat and stray rusty bolts fill the former floodplain in chaotic formlessness. The creek will stagger through the mess next summer, as inevitable as gravity, coffee-brown, streaked with waste engine oil. The willow thickets and flowers are gone, and with them the tattlers, thrushes, moose, jaunty warblers, peripatetic porcupines and other creatures great and small. The creek has no stoneflies, no grayling.

I can see, though, that life waits its chance nearby. An unmined side creek hosted a few grayling until freeze-up. The untouched slopes lean down into the naked valley, showering it with leaves, twigs and seeds. From a few piles of gravel, the roots of resilient willows, half-buried, thrust out. I look long at the soil at my feet and find the grounded parachutes of a startling number of fireweed seeds on the dirt, and I know that trillions of moss and lichen and fungus spores invisibly dust the claims. Ten years, fifty, a hundred: they will be back. (I think, smiling, of Loren Eisely's ambivalent welcome to the wild things invading the city in his poem "The Last Days").[38]

In recent years close to 700 placer mines have been operating each summer in Alaska, most of them in the Interior. Placer mining turns a streambed upside down. During mining and for a few years thereafter a heavily-mined creek looks like a war zone. Many aspects of placer mining are—or are intended to be—highly regulated. But does the law have a sound ecological basis? Is it enforced as though people in general really think it is a good one?

What We Know

Placer miners strip plants and unwanted soil (which, tellingly, is called overburden) from the area to be mined to get at the gold-bearing gravel, and from future campsites, airstrips, ditches, equipment yards and settling ponds. From this disturbed ground, and from the sluices and ore-sorting machinery, the water runs brown from its heavy load of sand, silt and muck. The nub of the environmental challenge, then, is to disturb as little ground as possible, to ready it for later ecologic healing, and to clean up the discharged water.

After the miners leave, coarse gravels typically comprise most of the surface mantle. Piles of overburden are scattered haphazardly across the floodplain, and the stream wanders vaguely through new channels destined to be changed with every spring freshet and summer rainstorm. The stabilizing of stream channels, streambanks and tailings, and recolonization by plants and animals may take scores of years or only a few, depending on site conditions. Years after mining a fairly complete and stable community may exist there; it may be very much like the premining one, or very different.

In any district placer mining affects only a small fraction of the land surface. Still, this fraction may be biologically crucial. As in hot deserts and semideserts, as well as in many temperate regions, interior Alaska's streamside environments support different, and often more productive, animal habitats than the surrounding terrain. The transition belt between aquatic and terrestrial systems is a place of high biological and physical traffic. Soils near active streams, on natural levees, often are better drained and warmer than upland or bog soils nearby, while nutrients are replenished by occasional floods. Water courses are natural pathways for wildlife on the move. In the Interior, riparian environments of alpine, subalpine, muskeg and even woodland areas often are dominated by a deciduous tall-shrub community of relatively high primary productivity. In summer many species of migratory birds flourish there, and beaver gather their building materials and yearly food supply. In winter the key vertebrate browsers of boreal communities—moose, snowshoe hares and ptarmigan—concentrate in those shrub habitats.

The physical processes of sedimentation, scouring, channel widening and delta formation that occur below heavily mined areas have been described in a rough way, but only a few of their consequences have been traced out. Fine sediments smother the eggs of fish, like salmon, that spawn in clean, gravelly stream bottoms. Arctic grayling, the most common and economically important resident fish of the Interior, strongly avoid turbid water while feeding. When forced to live in silty water, grayling suffer gill damage, grow slowly and eventually starve.[39] Any part of a normally clear stream that is actively mined or is downstream close to an active mine, as operations have been conducted, is lost for summer grayling habitat. Sediments and turbidity also affect the underpinnings of the entire stream ecosystem, the algae, whose productive activity is dramatically lowered by smothering sediments and light-reducing turbidity, and bottom-living insects, whose populations and diversity are reduced sharply downstream from placer mines.

Unanswered Questions

The main missing element is the time dimension. We simply have not looked carefully at enough abandoned mines, nor followed the continuing change at recent ones, to appreciate more than vaguely the dynamics of healing. One study near Fairbanks indicated that on the hillocks of gravel left several decades earlier by dredges, plants grew back far more quickly on remnant patches of fine-grained materials, and near water, than on drier and stonier surfaces.[40] On one tailing pile favorable sites may have grasses, shrubs and vigorous trees, while poor sites show only tentative recolonization by lichens and other drought-adapted pioneers. In the southwestern corner of the Interior (near Nyac, in the Kuskokwim Mountains), another student found the best populations of small mammals on unleveled tailings, in the shrubby vegetation ringing pools of water.[41] Except for those beginnings, the story is unfinished. What determines whether postmining communities of plants and animals will look and function like the ones existing before mining? What mainly determines how fast recolonization takes place? What happens to annual primary production, biomass accumulation and living diversity over time on abandoned mine surfaces? Is the spatial element—that is, the geometric relationship among mined and undisturbed areas—important in the pattern of postmining ecologic change? And, more pragmatically, what interests do humans have in the process, and how can we achieve them?

The time dimension is missing downstream, too. We don't know how long it takes sediment to flush through the system, or how often new episodes of siltation occur as abandoned mines are hit by floods. (During the 1967 rainstorm of which I have spoken, my family and I were marooned at a mining camp at Eagle Creek northeast of Fairbanks. The shallow creek became a torrent. After the water dropped we could see that the flood had eaten much farther into old mining ground than in undisturbed, willow-grown floodplains. Mining had stopped in that section of the creek at least 35 years before, but the overturned land was still unstable.) We don't know under what conditions toxic heavy metal ions, released from freshly broken rock surfaces by weathering and bacteria, become an ecologic or human problem. The whole complex question of changes in downstream channel morphology is unanswerable. Even while mining is going on upstream, we do not really know what happens to organic debris accumulations, detritus shredding by insects and decomposition by smaller creatures, the welfare of fish other than grayling and salmon, the fate of stream-dwelling birds like kingfishers and water ouzels that feed by sight on small fish and aquatic insects, or to the riparian songbirds whose supply of just-hatched aquatic insects is in jeopardy.

Current Regulation and Management

Given what we know and don't know, how is placer mining influenced by society's perception of its own many-sided interests? How and to what ends are placer miners regulated?

In my view, most Alaskans are, if not apathetic to mining problems, at least not deeply committed to their correction. They have no passion that comes from seeing a thing to be clearly wrong. A few react to the topsy-turvy valleys and dirtied waters with honest anger. An opposing few feel in their marrow the affront of an elusive frontier freedom beset by legal and official bounds.

Public interest in the environmental effects of mining is expressed at national and state levels through a system of laws and agencies that walk sometimes in tandem, sometimes one after another, and sometimes in different directions. Their overriding concern is water quality downstream, which comes down to meeting standards for settleable particles, suspended particles (turbidity) and toxic metals. Pieces, usually 1 micron or more in diameter, that settle out within an hour in an undisturbed container are called settleable solids. Those that don't result in long-lasting turbidity. The major toxic metals of concern are arsenic and mercury. The

1989 maximum settleable solids allowed by National Pollution Discharge Elimination System permits was 0.2 milliliters per liter of effluent. The federal permit defers to state standards for increased turbidity, limited to no more than 5 nephelometric turbidity units above background at the point of discharge. Measured at the same point, arsenic may not exceed 0.05 milligrams per liter above background.

The basic strategy for meeting those standards is to prevent solids from entering the stream. If there are no solids, then there will be no siltation, no turbidity and no excess heavy metals. Often, a reasonably practical method of intercepting most heavy material is to use settling ponds. The trouble is that although a series of well-designed and well-operated settling ponds will eliminate settleable solids, they will not reduce turbidity enough to meet standards. Reusing the water from the pond to sluice additional gravels—so that water is never or rarely discharged directly into the stream—reduces the turbidity problem. Miners argue, however, that recyling is too expensive, ruins machinery, or causes lower recovery of fine gold. Whether settling ponds and recycling are required (as they are now) and consistently used (today their use is spotty), all agree that reducing the use of water to strip overburden, and piling stripped materials in places safe from erosion, are fundamental mining practices that improve the quality of downstream waters. Recent field tests of chemical flocculents offer promise of reducing turbidity cheaply to very low levels.

All of this sounds straightforward, but the real world out on the creeks is a good deal more convoluted. Regulating agencies have reasonably clear legal mandates and standards to follow. The standards, however, are arbitrary. For one thing, they are fixed and uniform, whereas nature is dynamic and variable. For another, they are only sparsely backed by a foundation of empirical evidence showing how they relate to ecological impacts. Inevitably, miners, agencies and conservationists find ample room for argument in this wide hiatus. Likewise, when a regulatory agency tries to define what steps (out of many that an armchair ecologist and engineer could dream up) that a miner must take to conform with Best Practicable Technology or Best Available Technology, it finds itself with little engineering or cost data from actual operations.

State and federal regulatory agencies constantly hear contradictory messages from political leaders, itself evidence of sharply diverging views or plain bewilderment on the part of the interested publics. More than scientific or economic uncertainties, this confuses the picture of placer mine regulation. The purported lead agency, the Environmental Protection Agency, was buffeted by strong winds from opposite compass points in the late 1970s and early 1980s, sometimes toward stringent protectionism, other times toward abdication of basic environmental responsibilities. The state's two most involved agencies, the Alaska Department of Environmental Conservation and Department of Fish and Game, have been split horizontally into willing flesh at the operating level and weak spirit at the politically vulnerable leadership level. Local public sentiment, supporting miners, permitted near anarchy on the creeks for the ten years beginning about 1975. While a few miners experimented with settling ponds and water recycling, many urged overthrow of regulatory processes or a reclassification of streams to make convenient practices legal.

The long romance of the American public with frontier mining, and the presumed value of gold, have clashed head-on with a newer, strong, but not necessarily comprehensive environmentalism to produce an atmosphere in which bombast and posturing are seemingly better rewarded than reason. Science is not quite depauperate, but at the same time that political leaders decry the lack of answers, they withhold the money to obtain them. The interest of managers, and thus of most scientists, is narrowly focused on water quality in the short term. It urgently needs the complementary attention to systems ecology and the long term. Meanwhile, the creeks are muddied and the mined flood plains turned upside down each summer: the short-term fate of grayling and water ouzels fluctuates inversely with the price of gold.

Here and there a courageous field agent or a far-seeing miner shows the direction toward a less hostile relationship between ecology and economics, people and nature. In these individual acts of care and innovation, and even out of the smoke and noise of bitter political fighting, glimmerings can be seen of better times. Ten years after skyrocketing gold prices lured miners to the creeks, miners seemed able to throw back every effort to bring environmental sanity into the hills. Today they sound defensive: they plead harassment and economic woes. They are sitting at the bargaining table. Every year brings a new regulatory step and an increment of acceptance by miners of last year's proposals. Consistent improvement out on the creeks is still years away, but the determination to make those improvements is gathering force. Once the will is there, the way will clear.

Dilemmas in an Ancient Forest

The forests of southeastern Alaska that America bought in 1867 were both young and old, depending on the scale against which they were measured. Geologically, they were callow, the mountainous slopes they covered having been under glacial ice a few thousand years before. In human terms they were old, the individual trees growing undisturbed until, storm-battered and partly decayed, they fell in a gale or under the weight of wet snow in their third or fourth or eighth century of life. Only a few were cut by the scattering of Indian people who lived in the region, and even the rising demand for lumber for towns and canneries in the late 1800s and first half of the 20th century made slight inroads in the vast seaside forest.

Local demand for lumber didn't require logging on a big scale, and timber exports didn't become feasible until rising U.S. and world demand in the 1950s sent wood prices soaring. In that decade the United States Forest Service struck a deal with two Japanese-owned firms: if those firms built regional pulp and lumber mills so that manufacturing jobs would stay in Alaska, the Service would assure a 50-year supply of timber at favorable (often, it turned out, below market) prices. By the late 1950s large-scale logging, using clear-cut methods in which all trees in a marked area are felled the same season, had begun in southeast Alaska.

From the very first, the Forest Service was aware of its responsibilities not just to maintain timber volumes over time and to keep the mills supplied, but to be good stewards of land. In the 1950s and 1960s the research arm of the Service watched what happened to soils in logged areas, and developed general guidelines for timber sale layouts. Avoid steep slopes, they advised; avoid windy peninsulas where exposed trees at the edge of clear-cut will later blow down. Build logging roads with an eye to drainage and erosion, and raft logs in water deep enough to float them at low tide. Other rules showed concern for the effects of bad logging practices on salmon streams, not based on solid local studies, but on experience in the Pacific Northwest where timber and fishing interests had been arguing for a long time about jamming streams with debris, and about stream siltation and temperature changes in spawning and nursery streams after shading trees were cut. The Service also identified "steamer lanes" where logging would be done in patches hidden from the view of tourists aboard cruise ships and ferries.

Game managers were slower than fish managers to express anxiety over the effects of broad-scale clear-cut timbering. Many had been told as college students in the 1940s and 1950s that good forest management is good wildlife management: in particular, that logging keeps the forest young and productive of deer, other plant-eaters and their predators. It wasn't until biologists had watched deer in mild and severe winters for almost two decades that they realized the old ideas didn't hold in Alaska. Deer seemed to survive winters of deep and persistent snow better in the old forest than in the browse-rich clear-cuts. From 1970 to 1980 when the concern was raised in earnest by state biologists, close attention was given to the problem. The general situation now seems clear. Deer need shelter and easy movement in winter as much as they need food. Hoarding energy reserves, in fact, may be biologically smarter than looking for food. Stands of big old trees intercept a lot of snow, keeping snow shallow underneath, while occasional very small openings (often blowdowns) allow patches of browse to grow. A clear-cut, on the other hand, eliminates shelter even while it temporarily increases food supplies. A recent clear-cut is a feast for deer in mild winters but a baited trap in severe ones.[42] Furthermore, once the branches of young trees form a continuous green umbrella, about 25 to 30 years after logging, a period of about 200 years ensues during which the shaded forest understory is nearly barren.

One major reason that wildlife concerns loomed larger in the 1970s was that in 1969 the Forest Service signed a third major timber sale, involving 50 years and 8 billion board feet of timber on one million acres in northern southeast Alaska. Any complacency about wildlife being protected by the sheer size of the country disappeared with that contract. The same sale that the Service saw as a great step toward full use of timber resources aroused biologists, hunters, fishermen and wilderness enthusiasts to a pitch of concern.

Another event that fed these fears was the transfer of land to native corporations under the 1971 Alaska Native Land Claims Settlement Act. Land and trees that the Forest Service had counted on to help provide timber for three pulp and lumber mill complexes no longer were available. This meant that the Forest Service would have to meet its contractual obligations from a smaller stock of timber. There were immense pressures within the Service to say no to any proposal to keep its old growth forests uncut.

A conservationist lawsuit filed in 1970 to overturn the third major timber sale delayed construction of the required mills for several years. Meanwhile, pulp and timber prices declined dramatically throughout North America. In 1976 the company that bought the contract asked to be released from it. The Service, extremely sensitive to conservation interests that had just forced a major congressional revamping of Forest Service policy nationwide, agreed. There has been no serious talk of a third major sale in the decade and one-half since then.

With the passage of time biological concerns about logging in the cool rainforests of southeast Alaska have tended to move from specific problems to more general issues. Once worried about blockage of salmon spawning migrations and siltation of spawning beds, aquatic biologists began looking at the habitat requirements of all freshwater stages in the life of salmon, and of Dolly Varden charr, cutthroat trout and steelheads as well. From there it was inevitable that attention would shift to the invertebrate animals that are eaten by these fish. Now a few biologists are wondering how logging might be influencing the whole energy and nutrient budget of streams from their beginnings high above the logging shows to their meeting with the tides.

The same is true of wildlife concerns. The specifics of deer winter habitat, bald eagle nesting requirements, and brown bear-logging relations have broadened into questions on a bigger scale of time and space. If all the logging proposed by the Forest Service, Native firms and the State of Alaska actually were undertaken, how much old growth would be left near sea level in 20 years, or 50 or 100?

Where would it be, and would that distribution best fit human and wildlife needs? After old growth is cut, what kinds of environments are created by plant regrowth? What happens to wildlife because of the particular size, shape, spacing and cutting frequency in a given watershed? In the broadest sense, how will people and animals get along in a landscape dominated by chainsaw geometry?

Some of these questions were in the minds of people who took part in the great debate over the fate of Alaska's federal lands in the 1970s. To the extent that the issues were dealt with at all, however, the 1980 Alaska National Interest Lands Conservation Act used old-fashioned political allocation methods: so many acres dedicated to your interests, so many for mine, so many to be decided later. Congress wanted some wilderness in Tongass National Forest; it designated over 5 million acres, mostly in mountainous country not coveted by timber interests Congress wanted to maintain a commercial timber supply at least equal to that which loggers had cut in recent years, so it designated very little of the high-production timberlands as wilderness, committed itself to spend at least $40 million annually on new logging roads and intensive forest management (aimed toward maximum wood product yields), and committed the Forest Service to offer for sale at least 4.5 billion board feet of timber per decade. Congress wanted to protect as much of Admiralty Island's wildlife and recreational values as it could, so it established a wilderness area of 0.9 million acres on that gorgeous island. But Congress also let native corporations select lands on Admiralty and stopped the wilderness short of an important mineral deposit on the island's north end. While establishing the Misty Fiords Wilderness in extreme southern southeast Alaska, Congress excluded a very large molybdenum deposit and the lands necessary for access and mine development. And so it went, a political Monopoly game in which players bargained for properties as the roll of dice and their political chips permitted.

The trouble with political bargains is that they mirror current relative power better than current human needs, ecologic realities or the always-discounted future. Inevitably, they are unstable. By 1986 conservation groups had persuaded congressional representatives to introduce a Tongass Timber Reform Act, which eventually worked its way to the desk of the President and was signed

November 28, 1990. This act eliminates the freedom Tongass timber development money enjoyed from annual congressional review and appropriation; now each year's request will have to stand on its merits. It removes the specific target of 4.5 billion board feet of timber offered for sale from Tongass forests each decade, a provision that may not change the number of trees cut in the long run but discards a golden number that mesmerized contending forces in the southeast Alaska debate for ten years. The act changes details of the two, 50-year contracts (which have a dozen years or so to run), tipping the balance toward public needs and wildlife requirements while still assuring reasonable wood supply to industrial loggers. Commercial timber harvest is prohibited 100 feet of each of most streams on Tongass National Forest. Lastly, the act establishes six additional Wilderness Areas totalling 296,000 acres plus 12 areas, almost three-fourths of a million acres all together, that are to be kept roadless. In a letter to Alaska Region employees of the Forest Service, Michael Barton, Regional Forester, said "It is important that we put the divisiveness resulting from the heated debate behind us...Let's move ahead...," and Alaska's senior senator Ted Stevens said "After 19 years, it is time to seek peace."[43]

Either hope or exhaustion may indeed bring quiet to this Western front, but the deep differences in social beliefs and personal values that brought on the battle are still there. They are not hard to identify, but the processes of technical-professional resource management and national (or state) politics have tremendous difficulty in centering on them. Almost any proxy argument seems preferred: legal technicalities, weaknesses of scientific evidence, economic indicators, ecological disturbances. But solving the proxy problems does not affect the deeper disharmonies of values, it merely sets the stage for new symptoms to arise.

One well-entrenched social myth, at once pervasive and powerful, yet rarely spoken aloud or critically examined, is the notion of settling frontiers. Empty land is an affront and a challenge. Good public policy, the idea goes, assures that empty geography is filled with the right people doing accepted things. In southeast Alaska the Forest Service from Pinchot to Cliff understood its mission to involve the promotion of permanent settlement of southeast Alaska by people working in the timber and dependent industries. Hence the 50-year cutting contracts made contingent on building pulp and lumber mills in Alaska, against the grain of economic efficiency. Hence, too, the recurring justifications of weak forest use regulation on the grounds that stricter controls would threaten jobs, and the Service's reluctance to ask Congress to designate wilderness where big trees grow, or to decide itself to set lands aside administratively from logging.

I have no quarrel with that as a historical social choice. But at some point it is no longer useful to make decisions according to a policy designated to meet perceived needs of a bygone time. What we do now know of the environmental and economic *costs* of maintaining or expanding employment in the timber industry? Is it time to review the policy of requiring primary manufacture (processing) before export, rather than allowing shipment of logs in-the-round? Do we have an equal obligation to channel timber yields through small as well as big firms? Must timber management by the Forest Service pay its way in economic terms? If so, by what system of accounts do we judge present practice? Anticipating an idea I'll come to shortly, are we now at a point when we should rethink what is socially acceptable, economically feasible and ecologically sound?

The Forest Service in southeast Alaska has not escaped—some would say it has never struggled against—the strong predilection of American society and its government to judge proper policy on the basis of the output of things easily measured. Although Congress consciously asked the Service to supply intangible, largely unmarketed goods like fish and wildlife habitat, scenery and settings for esthetic and recreational experiences, it has given the Service few tools and little encouragement to do so, compared with money made available to cut timber. (The forestry profession has tried harder than political leaders, I think, both in tools and encouragement, to equalize attention given to marketed and unmarketed yields.) This is an old, almost trite, criticism—but true and important nonetheless.

The difficulty, I think, stems from two sources. One is that the things people hold dear simply are not as one-sidedly utilitarian as the values expressed in economic and many political transactions. Why did the little settlements of Tenakee Springs and Port Protection join suits in the mid-1970s to stop local timber sales and logging roads in Alaska's Panhandle? Why, in 1989, did residents of Hoonah force the Forest Service to reconsider and revise timber sales? It is, at heart, a matter of how the residents want to live.

The other problem is that we still have trouble measuring the utility of things, as opposed to their price. We feel the worth of many goods not exchanged for money. If we don't try to measure their value monetarily, we run a big risk of being ignored or given the leavings from the political bargaining table. If we do, then we risk diverting our own attention and that of everyone else from the thing itself—the river of clean water, the experience of solitude—to its surrogate, the dollar value assigned. Dollars homogenize values.

The problem the Forest Service faces goes even deeper. Southeast Alaska is very different from the rest of the state. Under other historic circumstances it might easily have become a separate state or part of British Columbia. That distinctive social, economic and ecologic region is almost swallowed by Tongass National Forest, except as the communities themselves exist on small enclaves of private and state land. Everything a southeast Alaskan wants to do on lands beyond the enclaves, singly or as a local or state government entity, must meet the approval of the Forest Service. Every action taken or rejected by that federal agency changes conditions of life for the Southeasterner. If the Forest Service lowers stumpage rates on timber sales, less money goes into local government through revenue sharing. If the Service lets helicopters ply between the cruise ship docks and the Juneau ice fields, people in town have their windows rattled. If a long-term timber sale leads to an additional pulp mill, more year-round mill jobs and seasonal wood jobs, but perhaps fewer tour guiding jobs, are available.

What this means is that the Service has a terrifying social responsibility without the authority to admit it, the means or traditions to carry it out, or the structure of accountability. The voters of Southeast Alaska don't elect the Regional Forester or even the Chief Forester in Washington, D.C. It may be that the Forest Service does as good a job in the situation as any agency could. But the task itself is impossible. No amount of local public participation in planning can obliterate the fact that the Service is answerable to national needs and policies, not purely local ones; or that its core mandate is natural resource management, not the determination of regional economics and community life.

A final and almost tangential comment seems worth noting, and that relates to the range of time scales of Forest Service planning. On the one hand the Service must respond to immediate demands for resources by its many clientele groups, so far as the land seems to allow. On the other hand it must plan to maintain some level of yield of renewable resources in perpetuity: in effect, it must play God. Between now and infinity lie the 5-year timber allotments, the 50-year timber sales, the 100-year rotations for spruce sawlogs, the commitments in timber harvesting rules to the notion that Sitka spruce will always be more valuable than western hemlock, the decisions to spend money killing alders now on the gamble that such an investment will pay off in a 20 percent shorter rotation period, the decisions to clear-cut in spite of a later probable 200-year hiatus in deer winter habitat—or its permanent loss—and so on. There really is no escape from these decisions or their consequences. The only hope, perhaps, is that resource reserves will always be greater than the level of demand we foster or attempt to meet. And that seems less likely with each passing year.

Plows North

In the bright strong sun of early May 1989, the hawk turns lazily on a pike-pole spruce snag thrust up from a bulldozed pile of dirt, roots and trees left by land-clearers nine years before. For a half-mile north and south, and for an eighth of a mile east and west, the wet and disheveled fields extend to the irregular edge of spruce woods. Like several hundred of its fellow rough-legged hawks and companion red-tails, the bird has stopped to hunt voles on this new farm near Delta, Alaska, almost at the end of its spiralling flight from its winter quarters, perhaps in Arizona. Voles are scurrying everywhere, their runways and shallow tunnels too wet for comfort and the green tangle they will hide in next month now only a hope. The land clearing and fitful pioneer farming of the past ten years have been good for the voles and hawks, good for the gawky sandhill cranes and elegant geese that glean the abundant waste grain and new spring blades.

Whether the work has been worth the effort for those who did it or those who helped pay for it remains a question this spring morning. The whole country looks uncertain. Some fields were harvested last fall and are ready for planting. Others are thick with barley so flattened by August's bison and September's snows that they couldn't be cut. That work may or may not be done this spring. Still other fields were cleared, planted and abandoned all in the space of five years. Willows and yarrow are invading fast. The farmsteads—and it is an act of faith to call them that—are just as tentative. Some few are nice houses, the kind you pay $150,000 for on the subdivided hem of Fairbanks, looking grand on their half-acre lawns. Here, they look particularly alien. What jars the eye and then the mind is that the land around them clearly cannot support their implied luxury. Of a holding of perhaps 700 acres, only 10 are cleared. Five are in hay for the two Arabians nickering in the corral. The house and its people are perched on the ground, but their sustenance comes from away: the North Slope oil fields, perhaps. Whether they will ever take local root is anyone's guess.

Other farmsteads were accumulated rather than built and have been worn out since birth. An old trailer jacked onto stumps, scattered fuel barrels and propane tanks, second-hand sheds, several three-wheelers and snowmobiles and one ancient rusty "cat," a small woodpile beside a tangled mound of tumbled trees: a long way from here to the settled farms of Wendell Berry's Kentucky.

Here and there you find a farm that reflects what the land—the best of it—and a serious farmer—the best of them—can do in partnership in this place. There are some long-term questions to be answered about how long the soil will last and whether world or local markets will allow a profit margin, but for now, only a few years from the first conversion from spruce and aspen to barley and rapeseed, the dedication, care and experience of these farmers is evident on the land.

The Delta Barley Project was the political outcome of efforts by organized Alaska farmers to push agriculture into the mainstream of resource development activity. All but forgotten in the hardscrabble decade after statehood, farming was totally eclipsed in the early 1970s not only by tourism, fisheries, mining, logging and practically every other work of traditional interest to Alaskans, but by North Slope oil activities and pipeline construction. Looking for something spectacular to make agriculture more prominent, a few farmers and legislators—helped by several agency heads and leaders in academic research in local agriculture—conceived the idea of growing feed grains for export to the Orient and to stimulate a local pork- and beef-raising industry. World prices for feed grains were up, and we knew that cool-climate grains of food quality could be grown in interior Alaska: why not raise a lot, and export them?

And thus, a project was born. The state would make land available, selling at low prices the right to farm. It would loan farmers the money for capital improvements, equipment, seed and fertilizer. It would build roads into the Delta Project area, subsidize power distribution lines, help put up a grain drying and storage facility, build a loading terminal at a deepwater port on the Gulf of Alaska, pave the way for foreign buyers to sign contracts with farms, and support the price farmers received for their grain. When the time was ripe, the state would loan (at interest rates well below market) money to help hog farmers, beef raisers and slaughter-houses get going.[44]

The political timing was good, and the state invested a lot of money in the idea. Unofficial figures put the outlay at $30 million by 1980, or about $1 million per farm, but the state has kept no separate account of Project costs. While the state was spending over a billion oil dollars each year on a gourmet's menu of capital projects, it seemed reasonable that agriculture should get its share. Land was surveyed, cleared, planted. By 1983 almost one third of the 60,000 acres in the original Project area were being planted each spring. The State was on the brink of adding another 55,000 acres neighboring the Delta Project, and residents of Nenana, a town 100 miles west of Delta, were demanding a similar but bigger project in their area. Trying to regain a competitive scale of milk production after the near collapse of the dairy industry in southcentral Alaska in the 1970s, the State sold 15,000 acres of land for dairying and general agriculture just north of Anchorage in 1981. Policy analysts estimate that the State will subsidize each of 20 farms there to the tune of $1.3 million.

The view from 1991 isn't so hopeful. Political support for agricultural expansion projects has plunged with the fall of oil revenues: farmers simply don't weigh enough. Production of grains in the original demonstration project area at Delta is stuttering along at levels below those achieved early in the Project, which were themselves far below those supposedly necessary to support grain exports. Some Delta farmers aren't farming anymore, some have gone bankrupt, almost all are deeply in debt, a few are doing well with specialty crops like grass seed. The area's only slaughterhouse has gone out of business, and the sole new proposal for slaughter facilities has been abandoned. The Delta expansion is on indefinite hold, and support for a Nenana project is vanishingly small. One of a handful of dairies in southcentral Alaska has essentially been taken over and run by the state.

Curiously, farmers who sought low-interest loans from the state government from 1978-80 to clear, plant and harvest grains, by 1987 were applying to the federal government for money *not* to plant. And the Soil Conservation Service, which for 10 years had been tacitly certifying that Delta land is farmable and had helped develop techniques for low-erosion cultivation, is now busily certifying that the land is highly erodible and qualifies for the new federal Conservation Reserve farm program to reduce production and slow erosion.

To make too much of these problems is as bad a mistake as to ignore them. Growing crops in Alaska is both a possibility—a certainty, in fact—and a good idea. A half-million people can't gather food from Alaska's not-too-fruitful wildlands even if they all wanted to, and it doesn't seem wise to import everything from Outside. Foraging and grocery imports are necessary but not sufficient: it takes three legs to hold up Alaska's belly, and the third leg is growing things ourselves. Beyond that, there may be a few agricultural products—in addition to reindeer antlers—we can export profitably.

The problem is our impatience. We seem to want to make a killing before we know how to earn a living. The frustration of local farmers in the early 1970s, who felt isolated, ignored and held down in their prospects while colleagues in Iowa reaped bonanzas in corn and real estate, was translated into a hasty, poorly conceived political project. The best we could do with our imagination was to see how closely we could mimic the results of 100 years of struggle by Midwest farmers without going through the process of getting there. Money compresses time: spend enough, we thought, and you can make anything happen overnight. We did, but it didn't.

At Delta, in spite of government's single-minded fixation on a one-crop monoculture of feed-grain barley, the farmers themselves angled toward polyculture. They raised barley for saddle horses and pigs. They sold straw to Alyeska Pipeline Service Company for use in case of oil spills. They produced hay for the pets of local trailriders. They grow pure stands of grass for quality commercial seed, weeds not having gotten a foothold yet in this pioneer area. Pure balkiness might explain the rebellion, or an unspoken disbelief in the promises of an export market, or a reaction to repeated "unusual" years in the early 1980s when, for one reason or several, barley crops failed. Or, perhaps, the reason is simple experimentation and opportunism.

In the rude awakening of the mid-1980s, when the honeymoon was over for the Delta Project and the possibility was admitted that the experiment might fail, there have been signs of a willingness to look for more appropriate agricultures. There is a great deal of room for such creativity, given the immense cultural and ecologic diversity of Alaska.

First and most obviously, perhaps, northern agricultures need to be culturally valid. One of the vexing problems of agribusiness in other states is that as profitable as big farms can be, we are still not thoroughly convinced that the profit is worth what we pay for it: the breakup of farm families, dependence on banks, corporate takeovers, and the costs of monoculture. And the difficulties experienced in trying to inject machine-driven, energy-dependent and capital-devouring agriculture into Third World countries are well known. Appropriate agricultures for rural Alaskans who have subsistence-based cultural traditions should fit with the habits of self-sufficiency, small-group sharing and seasonal gathering of wild food; with their diet preferences; and with the scales of capital formation and cash flow typical of those regions. There may be no model for such agricultures anywhere in North America today. Something brand new can evolve.

In suburban and rural (as opposed to hinterland or "bush") Alaska, there is such a mix of people and accepted values, and so rapid a change in them, that the most we can say is that conscious choices about appropriateness need to be made. If enough people want large-farm, capital-intensive, high-risk agricultures, then there is certainly space to try them. (As with all other agricultural styles, our social decision is how much, both economically and ecologically, we are willing to pay to start and possibly to sustain them.) For others, the family farm, mixing animal and plant culture, and coupled with varying amounts of income earned off the farm, seems to be the model of choice. The process of adapting this general style to northern realities has begun and will continue, with or without help from governments and scientists.

The important thing is that the values cherished in Alaskan societies should be the underlying foundation of development in agriculture.

It is tempting to state that appropriate agricultures must be economically valid, too, but I admit to a little uncertainty on this point. If we think of farming solely as an orthodox economic enterprise, the private and public money returned from farming should exceed the private and public money put into it, by an amount acceptable to investors. Fair enough, as long as the private and public profits each come in proportion to the investments. We can even allow a bit of a caveat: that some of the returns come as meat and vegetables consumed in the farm home and other in-kind values, not cash. But agriculture is not always seen as a purely economic enterprise. If a farmer is willing to accept less-than-maximum profits to gain a pleasurable kind of life, and has the savings or other income to do so, should we prohibit it, scoff at it or ignore it? Or does it move into that large class of activities that are not usefully measured by profit: religion, education, recreation, art, science and marriage?

In its more restricted sense as an economic enterprise, agriculture should meet several common-sense criteria. Profitability should be a reasonable expectation both in the short run (a few years) and in the long run (several generations). That is, the range of crops that are growable should be such that, in most years, a prudent person would expect to find at least one that can be harvested and sold at a gain, or eaten. There should be enough income, before profit-taking, to maintain the physical and biotic systems on which the farm depends: primarily the home, utility buildings and soil. Somewhere in the flow of revenues there should be enough to support infrastructure costs (farm roads, power distribution systems, schools), research, and the costs of unwanted side effects of farming, such as environmentally broadcast chemicals, wildlife habitat losses and similar externalities. Finally, considering Alaska's location, there may be an economic advantage in farm systems in which dependence on chemical, machinery and other imports from

Outside is pared to the bone, not just in recognition of transportation surcharges, but also as insurance against the unexpected—like longshoremen's strikes.

Ecologic appropriateness is a third major dimension of agricultural suitability. Obviously, suitable crops are those that can be harvested in most years, not just in the mildest. Crops adapted to no-till systems will promote better soil care but may require more herbicides. Crops that are soil-builders or are thrifty in their use of nitrogen, phosphorus and other nutrients will be best in the North's nutrient-starved soils. Farming that promotes on-farm decomposition and nutrient recycling is wiser and cheaper than linear, import-export systems. Very likely, the same advantages crop diversity has in temperate North America would apply in subarctic zones as well.

In sum, appropriate agricultures are being and will be devised for Alaska. Creativity in adaptive farming practices, and the willingness to think of agriculture in contexts broader and richer in human content than orthodox economics, are key ingredients to success. So far, neither government-supported research nor politically spawned development projects have been notable for those traits.

Gold, grain and slow-grown spruce: we are learning. At Eagle, Mammoth, Crooked, and Golddust creeks, the water still runs with unwanted silt in summer, but not as much as a decade ago. On the creeks in 1978 it was "Damn the rules and damn the distant bureaucrats!" but now, almost unbelievably, mining is moving out of its rough-hewn history toward a more responsible future. In southeast Alaska's green rainforests, we are still gridlocked on crucial issues of forest management, but at least there is an emerging sense of the uniqueness and value of ancient forest forms and systems. And in agriculture we might—just might—have learned the dangers of public force-feeding of a young enterprise. After trying to leap from feeding grain to a few local riding horses to competing with Kansas for Japanese contracts, we may now be willing to learn to walk before we fly.

Each of the examples I selected shows that today's resource stewards are well aware that resources form interacting clusters and that these dynamic relations can last a long time. If a resource manager wades into one of these clusters with the notion that one resource dominates all others, someone in our pluralistic society is sure to attempt his or her re-education—perhaps in court. And so pioneer farming is seen, quite properly, as a process of broad ecologic change; placer mining is expected to matter to wildlife enthusiasts, anglers and downstream water users; and the felling of a centuries-old hemlock is a concern of an entire regional biotic and human community.

It is also true, I think, that we accept some responsibility for investing in good stewardship. As threatened as federal and state program budgets for resource management have been in the 1980s, by far the majority have survived—some barely clinging, some inching ahead. However, it is also true that public expenditures for extraction or production phases of resource use often far out-weigh investments in maintaining resource flows over the long haul, insuring against overexploitation, and protecting the soil and water base that underlies renewable resource yields. We haven't yet won the battle against myopia.

5
Fitting into the Country

Enduring relationships between people and the rest of nature seem possible along a wide spectrum of life styles and environmental conditions. At one end is the venerable tribal lifeway in which people change their surroundings very little. Local ecologic processes are still controlled almost entirely by weather, landform and local energy and nutrient cycles. People live mainly by harvesting and harnessing endemic renewable resources like trees, wild food plants, wildlife and the energy of river, sun and wind. Tools and machinery are simple, trade and long distance travel are less important than internal economies and movements. Human numbers move temporarily up and down around a long-term average reflecting local biological productivity (which varies and may be affected by human presence) and the slow evolution of technology.

In some extreme science-fiction alternative, we can imagine a human society living in a world in which everything from atmosphere to ocean shows a heavy human footprint, a world where landscapes have become culturescapes. Plants and animals are genetically engineered from a universal basal stock of DNA. Minerals and petroleum are mined extensively and recycled intensively. Food is grown in production chambers and much of what people eat is constructed from a standard suite of protein and amino acid building blocks derived from algae or artificial leaves. Human numbers are limited by necessary ceilings on pollutant outputs and inescapable inefficiencies in recycling of energy and materials. Nature is everywhere, but everywhere transformed. This is the future Bill McKibben ("The End of Nature"[45]) persuasively insists is here already—a place where everything is, perforce, still of nature—that is, made of matter and energy—but so rearranged that nothing natural still exists in the sense familiar since the birth of humankind.

All existing lifeways lie between those extremes. There are hunter-gatherers—probably over 100 million worldwide, counting those who also grow supplemental crops—but almost without exception they have adopted some modern machinery and require cash that they get by selling goods or working part-time in the industrial economy. There are stockherders with very simple technologies and low material consumption. There are farmers in the modern style, wonderfully productive but increasingly vulnerable to the whimsies of the very factors that made their life possible: world trade, mobile capital and high technology. And there are the hundreds of millions of modern urbanites and suburbanites whose lives and surroundings seem, deceptively, to have drifted far from nature's domain.

Are these different lifeways durable? Hunter-gatherer cultures hold the record so far, having answered a range of human needs— more accurately, having defined humanity—for over a million years. There isn't any doubt, however, that although the broad style is ancient and highly durable, countless practitioner communities in particular times and places have failed and vanished. The same is true of herding and traditional farming cultures, known to have at least a 10,000-year history. Time after time, herding and farming practices have failed to meet demands either of ecological fitness or of human population growth, and have crumbled. As for our industrial or postindustrial lifeway, too new to have proven durability by long survival, I'm convinced, as are many others, that our early and current patterns of thought and action are insupportably destructive.

Walter Firey[46] presented a useful model of the broad requirements of durable interactions between people and environment in

"Man, Mind, and Land." He concludes that the most enduring resource practices are at the same time profitable, socially acceptable and ecologically supportable (Figure 6). At any given moment many practices meet only one or two of those criteria and are inherently unstable. Resource behaviors that meet all three tests are the most sustainable.

The most interesting aspect of Firey's schematic is its inherent dynamism. If you think about practices that are near the edge of any circle—the sphere defining gainful practices, for instance—you realize that there are both centrifugal and centripetal forces at play. Just outside Fairbanks' city limits is a hill with substantial gold peppered through its schist and quartz. Around its eroding skirts placer (gravel washing) operations have continued (with ups and downs, of course) for three-fourths of a century. A few early miners sank shafts into the bedrock of the hill's flanks and shoulders, trying to find ore rich enough to earn a profit after blasting, crushing and milling. Those practices had limited and short-lived success. In the last few years, however, a company using huge earthmovers has stripped one hip of the hill and used cyanide leaching techniques to free the almost-microscopic gold from the country rock. Even at gold prices less than one-half of their peak in the late 1970s, this latter practice seems profitable. On this one mineralized hill, therefore, are examples of a resource practice that has been quite durable, another that fell below the profit threshold and was abandoned, and a third that at least for now is expanding the reach of economical mining.

The interactive dynamics of gainfulness, acceptability and ecological soundness are part of the history of that same high, rounded hill. Early placer operations washed millions of cubic yards of silt out of the floodplains of neighboring creeks; never ecologically sound, this activity was, nevertheless, acceptable to local society for decades. Placer mining became a hotly debated issue in the 1970s, and today the old methods are illegal, replaced by practices that more closely meet today's stricter environmental expectations. The new strip mine, profitable as it may be, still jostles residents who dislike the din of machinery, abhor the spreading scar on the green hill, and doubt the firm's ability to keep the cyanide from poisoning downslope soil and water supplies.

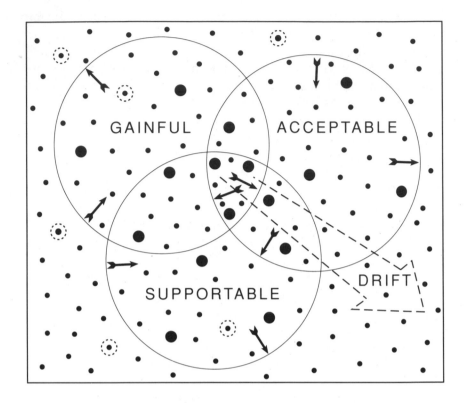

Figure 6. A model of the dynamics of resource practices (from Firey, 1960). Within the universe of all possible resource practices (field of dots), some are socially acceptable, some are ecologically supportable, and some are economically gainful (large circles). These arenas overlap somewht, but not entirely. Some resource practices have been abandoned (circled dots), others are in use (large dots). Forces (small arrows) continually act to expand or contract the area of the circles; examples are technological change, ecological change, and changes in human values. The broad arrow suggests that there may be trends in all spheres more of less simultaneously.

Climbing up for a broader perspective of Firey's model, we can see that the whole three-circle group drifts slowly through the field of potential resource practices. The reasons are many. Social mores change, sometimes slowly and sometimes quickly. The virtual disappearance of slavery, the rising standards of treatment of wage-earners and the burgeoning animal rights movement are three of many examples. A study of human history also suggests that there are cultures that are made stable by the strength of tradition, where resource practices change only slowly. In others, like ours, change is wanted for its own sake. Technological changes constantly make the impossible, possible; at the same time they force abandonment of once-widespread resource uses. And, as we are now well aware, nature itself is far from static. Regional climates, the shape and position of continents, the level of the oceans, the character of regional vegetation and the composition of Earth's atmosphere all are, on different time scales, variable rather than fixed factors.

It is in the context of these three sets of criteria for resource practices, and their dynamic relationships, that I want to sketch some guidelines for individual and communal behavior in the North. Some are mainly ecological, some deal with economics and others incorporate tests of social acceptability. However, the three criteria inevitably intermix, and I haven't tried to force them into rigid and separate molds.

Least Disturbance

The principle of *least disturbance* recognizes the depth of our ignorance. We are newcomers with southern roots, to whom northern environments are strange; or we are indigenous people to whom the values, institutions and technologies of modern life are shockingly unfamiliar. Human experience teaches that the greater the disturbance the more likely that big, unexpected and unwanted effects will ripple outward through time and space. The less we know about a landscape, the more dramatic and frequent the surprises. There is wisdom in seeking our ends with the least change to existing environments, and wisdom in asking frequently whether our ends and our means are still sound.

After Fairbanks flooded in 1967 and the worst was over, town leaders asked the Corps of Engineers to do something. Traditionally, that "something" would have been a dam and reservoir, the impoundment being kept nearly empty so that flood water could be caught and released slowly. Instead, the Corps chose a gate-and-diversion system. Until a major flood occurs, the Chena River flows unimpeded. During a flood the gates can be lowered to skim off the crest. The diverted water follows a dike just upstream from town until it flows into a larger adjacent river, the Tanana, bypassing the city. Building the dike and flowage way did dramatically change a forest mosaic into a grassy sward, and the absence of floods below the gates will eventually change the dynamics of the lower few miles of the Chena. Together, however, these changes are much less than those accompanying a dam and reservoir.

Many things we do in the course of daily living require that we maintain a gradient between human-altered systems and surrounding ambient conditions. The cost of maintaining the gradient is often proportional to its magnitude. A lawn of Kentucky bluegrass is harder to maintain in the subarctic than a patchwork of local plant pioneers. As my wife will testify from her ongoing war with dandelions, the purer the stand of grass you demand, the greater the effort to keep weeds out.

Similarly, government efforts to put out wildfires in remote areas of fire-prone interior Alaska are costly in proportion to how many we struggle to quench. In the 1950s and 1960s, federal fire fighters acted as if all wildfires should be put out and that eventually they could be. Over time both notions have proved wrong. Fire is a fact of Interior ecological functioning, and much of the region's diversity of species and productivity of plants and animals is due to the existence of wildfires. And the costs of full suppression of every fire are obviously too big to be borne, absent a huge surplus of oil money. (In 1990, by way of illustration, the state government had appropriated $3 million for fire suppression. Ten days into the fiscal year, in a hot July, they had spent $7 million.) Contemporary policy cautiously admits the benefits of fires and the budgetary impossibility of putting all of them out. Increasingly, agencies rely on a case-by-case evaluation of each fire in relation to a previous agreement about land management strategies, to decide whether and how much effort will be spent on controlling the fire. At least this is the official bureaucratic line: when fires actually flare up, the entrenched urge to put them out seems irresistible. Whether the progressive spread of cabins, settlements and other investments through Alaska's fire-region hinterlands will force us back to a policy of complete suppression, only time will tell. The point is only that our view of fire now includes a jot of biotic common sense, and our ideas of fire management are more consistent with the principle of least disturbance.

The principle of least disturbance, like others to be discussed, cannot be followed slavishly. Meeting genuine and important human needs almost always nudges or pushes environmental elements around. If animals and plants and natural systems have rights and are worthy of respect, so do and so are we. The principle clearly refers to a relative scale: it asks that we carefully question whether the action is necessary. If the answer is yes, it asks that we choose a way that results in as little environmental disturbance as possible.

Also, the idea of disturbance is tricky. Clear-cutting is a major disturbance of southeast Alaska's ancient rainforests, but is vastly less so in the Interior. Undisturbed by people, southeast Alaska's ecosystems are the result of venerable age, and depend on the decay and replacement processes and the complex niches everywhere from soil to treetops typical of ancient woodlands. Clear-cutting upsets that stability and replaces it with a perpetual cycle from youth to middle age back to youth again that is beyond the adaptive experience of many of the region's plants and animals. In the boreal forest, in contrast, animals and plants are well adjusted to the more transient life of pioneering and the rapid flux of nutrients, soil temperatures and soil moisture resulting from fires. Clear-cutting partly mimics that pioneer-demanding environment, and hence is a change that is often within the adaptive capacities of local biota.

Less Than The Most

Throughout American history the idea of getting the most out of natural resources—maximum extraction of minerals, maximum yields of renewable resources, maximum settlement of land—has pervaded our whole approach to local, state, regional and national development. One of the most common justifications for conquering or displacing native Americans was that they weren't using the continent's resources fully. Scientist-philosopher Alfred North Whitehead expressed the belief clearly in his book "Science and the Modern World" when he said:

For example, the North American Indians accepted their environment, with the result that a scanty population barely succeeded in maintaining themselves over the whole continent. The European races when they arrived on the same continent pursued an opposite policy. They at once cooperated in modifying their environment. The result is that a population more than twenty times that of the Indian population now occupies the same territory, and the continent is not yet full.[47]

To Puritans in early New England, land use meant land ownership, and ownership was critically evidenced by fences; Indians built no fences, hence must neither own nor use the land—which was thus free for the taking. A similar notion justified the displacement of trappers by ranchers, of open range by fenced range and farming, and of farming by suburbia—though somewhere in that sequence the marketplace became the acknowledged arbiter of decisions that once were debated on principle. The commercial frontier marched into Alaska by the same drum, and in 1956 the newly written constitution for the not-yet State of Alaska enshrined maximum use as a fundamental principle for the future:

It is the policy of the State to encourage the settlement of its land and the development of its resources by making them available for maximum use consistent with the public interest.[48]

The qualifier, "consistent with the public interest," was meant even then to be a powerful tool for preventing outrageous exploitation, violation of public and private rights and excessive waste. Later sections qualified and balanced the idea of maximum with clear statements requiring sustainable yield principles in managing fish, forests, wildlife and water, and authorizing reservation of sites of natural beauty and historic interest from public domain so they can be protected in perpetuity. Nevertheless, the politics of the pseudofrontier has shouted the slogans of use and development and whispered the words of restraint.

Unfortunately, "maximum" as a lodestar for management inevitably leads to brinkmanship. Examples are everywhere, from the king crab fisheries of the Aleutians and Kodiak Island to the caribou stocks of the Nelchina Basin; from the polluted air of Juneau's subdivided Mendenhall Valley to the poisoned cove into which Ketchikan's pulp mill dumps its wastes. In a sense these are merely mistakes, and human systems can't be perfect. But they are the kind of mistakes that are absolutely predictable in a strategy geared to maximum output. The dangers of going-for-broke are especially sharp in the North. Here, supplies of water, wildlife and fish naturally vary over a broad range from year to year. Crucial environmental conditions such as snow depths, sea ice distribution, summer drought and winter air inversions can't be predicted early enough for managers to prepare for them. Investments in resource extraction structures, machines and labor, being relatively costly, demand maximum profit-taking. And management effort usually is spread too thinly to permit control at the brink.

It seems logical and far more sensible, then, to pull back from the precipice of maximization and to reorient management toward yields that are less than the hopefully calculated limit. Although there are disadvantages to this principle of submaximization, especially if you consider an unused thing to be a waste, the advantages are compelling. Management becomes less costly because stocks and environmental conditions don't have to be watched so fearfully, and enforcement costs decline. Managing at the brink requires informational exactness: maximum yield, high-tech management and high-tech science go hand in hand. The consequence is that managers and their technicians become remote and inaccessible to the general citizen and their political leaders. Resource users save in a less-than-maximum system by encountering fewer years of enforced vacations while overexploited stocks recover, and they will run fewer risks of overcapitalization. In more general ecological terms, the risks of soil depletion, air and water degradation, and nutrient depletion (by continued removal of timber, agricultural crops and even high-biomass animal stocks such as salmon and pollock) will be less.

Tailor To The Country

A third guide to human behavior in the North is to *adapt to local conditions*—an idea so simple and commonsensical as to be trite. Yet, almost every institution we deal with in the North, whether of our own creation or not, finds it easier to produce uniform products than flexibly tailored ones.

Uniform technologies and systems of behavior control beguile us by their real or imagined efficiencies of scale and standardization. Architects, engineers and builders still prefer and use broadly standardized designs and materials rather than work out locally adaptive variants. We still attempt to control whole constellations of environmental degradation with uniform regulations. Schools and utility systems for small rural communities in the North still are conceived and built as scaled-down versions of urban, temperate-zone counterparts.

Going against the grain of global standardization is hard. The Alaska market for custom-tailored materials and hardware is small almost by definition. The constant influx of newcomers, many of whom are in the North only on short-term shifts, dilutes the interest in adaptiveness. Every level of governance seems reluctant to create flexible management regimes and standards, perhaps because flexibility implies decentralizing power. Logically, if the scale of environmental variety to which you need to respond is local, or at least regional, then the small-scale governments such as municipalities and boroughs (counties) are where you would look for appropriate scopes of decision. Unfortunately for this perspective it is the larger units of government, state and national, that have the critical jurisdictions. It is those governments that, in Alaska, own almost 90 percent of the land. It is they who set standards of air and water quality, land reclamation, forest practices, wildlife management, soil erosion control, environmental chemical usage and highway construction. It is they who set operational stipulations in oil and mineral leases and most timber-cutting contracts. It is federal and state governments whose grant and loan programs tend to standardize everything from energy retrofitting to park architecture.

I do not advocate wholesale transference of decision making to the local level to gain the benefits of local adaptabilities. The disadvantages of local control led us toward centralized control decades ago. The right balance hasn't been found yet, however, between the potential chaos of locally set management and regulatory systems and the unwieldiness of distant and broad-scale standards. It has become fairly common for a higher level of government to allow a lower one to take over an environmental management process like surface mine reclamation and water quality management under the spreading umbrella of uniform general standards. This is useful, and should be done more often. We could go much farther toward allowing smaller-scale decisions about how to meet performance standards. We should encourage local innovations in problem solving much more openly and actively than we now usually do. Briefly in the mid-1970s, the State of Alaska had a grant program to promote creative, small-scale energy conservation and production systems. Briefly, a decade later, the State gave grants to placer gold miners for developing more efficient, water-conserving and environmentally gentler mining methods. Those kinds of programs could be applied much more broadly. The State's new (1988) Office of Science and Technology seems to be such an attempt, and I wish it well.

Stay Simple, Stay Flexible

One result of this approach would be the invention of less costly technologies than currently are in use, because the inventors would be highly motivated toward *simplicity*. Likewise, they would want to solve specific problems as they occur in real, locally varying environments instead of the statistically conceived but chimerical average conditions that broad regulatory and political processes respond to. Simple technologies more often take advantage of local conditions instead of overriding them. They are more easily maintained and less subject to the whims and schedules of distant specialists. They cost less and often are more moveable.

I have mentioned *flexibility* as an attribute found in many northern plants and animals and highly valuable in human activities in the North. This essential flexibility is gained in different ways by living things: varying reproduction rates, inherent genetic variability, ability to escape periodic hard times, omnivory, food storage habits and so on. Humans can mimic all of these. In the North, for

reasons both ecologic and economic, a durable society will value
and fine-tune the generalist strategy (for individuals and corporations), reward low fixed investment and high efficiency, develop
systems of temporary settlement for short-term resource extrac-
tions, and create savings systems so that some profits during good
times are held over for use in hard times.

Resource bases and profit margins wax and wane rapidly in the
North both because of the region's fundamental ecological character and because of its economic remoteness and marginality. To
build conservatively, then, would seem a valuable economic principle consistent with the idea of flexibility. (My attention here is
mainly on public-sector investments, under the assumption that
the profit motives of private resource corporations usually will
lead them, individually if not collectively, toward a least-cost
approach to capital investment. This assumption breaks down
when many private investors compete for access to a common
resource, such as marine fish or underground petroleum.) In
Alaska neither the public nor political leaders ever seem willing to
admit that any given major project will be less grand and durable
in reality than in anticipation. Part of a strategy of sustainable
development should be to have the least fixed investment to
sustain through cyclic downturns.

The Sustainable Business

In any society interested in its own long-term continuance, there
would be a natural emphasis on economic activity that can be
sustained for a long time and that pays its own way.

Sustainability doesn't mean that every firm must survive, or that
the activity must use a constant technology or operate within
unchanging tax, regulatory or other institutional structures.
Sustainability does not require strict stability, either. Ups and
downs in overall economic volumes and profits in a given sector
are inevitable. What it does mean is that those fluctuations stay
within bounds (especially the lower bound of failure) that permit
society to plan on continued contributions from the arena of
activity. Activities that use renewable natural resources like flowing water, soil, forest lands, scenic attractions and so on, must not
erode their own base of natural capital or (as could occur if timber

production led to habitat deterioration for salmon, for example) the foundation of other economic activity.

The question of paying your own way, seemingly so obvious, raises some knotted problems. In the pure sense of the term, a firm would be paying its own way if it showed a profit without

1) requiring government subsidies,
2) eroding its own resource base,
3) causing uncompensated costs to other people in the form of bad health, lost amenities or opportunities, or
4) building up environmental debts such as polluted soil, air or water that eventually have to be cleaned up at public cost or allowed to deteriorate other resource yields and lower our living standards.

In some ways the condition least likely to be met is the first. In our economy everybody pays for subsidies to others and receives some in return. The subsidy system is so complicated that it is hopeless to try to figure out who is a net winner and who a net loser. More important, perhaps, although the subsidy labyrinth might be made more rational, it can't and probably shouldn't ever be eliminated—an acknowledgment of inherent flaws in capitalistic market economies as well as a bow to political reality.

Alaska is as thoroughly caught up in the maze as any other state. In addition, we have a number of perceptions—if not real conditions—that make subsidies even more common and likely more crucial as well, at least to some enterprises. For one thing the enormous prestatehood support of Alaska's fledgling market economy by the older states, justified as an investment to give the child a chance to mature, has declined but not disappeared. Because we have a comparatively callow and unformed infrastructure, because our population and hence market scale is small, and because imported goods are so costly, government subsidies are argued to be needed to start enterprises that, eventually, will be able to stand alone. Another simple and powerful reason is that oil revenues have given us (or, more accurately, did give us for a short time between 1979 and 1984), the means to indulge in generous subsidy policies.

What this means is that the Alaskan private economy is more public than we like to admit, more supported by a web of subsidies than perhaps any other state. "Paying our own way" is a bit of theory as yet untested in Alaska on any significant scale. We don't really know which of our current enterprises is sustainable. Since neither major fountain from which these subsidies flow—the federal government and oil dollars—is inexhaustible, we know the test will come some day.

The subject of externalities—the ripple effects of economic activities, both positive and negative, that are not accounted for in the profit-loss ledger of the firm—is huge and not necessary to talk about here. As long as ill effects on people and environments remain outside the accounts of benefits and costs, and outside of the responsibilities of the causing party, we won't know whether the enterprise is truly paying its way, hence truly sustainable.

A word about compensation for externalities is worthwhile. From an economic standpoint it may be enough if human winners pay human losers for costs incurred; for example, if upstream placer miners pay subsistence fishermen an acceptable amount of cash for lost fishing opportunities. One of the biggest issues of the hectic months after the *Exxon Valdez* ran aground in March 1989 was who would be compensation for what losses, and by whom. From an ecologist's standpoint, however, the aquatic ecosystem is still being impoverished. The transaction has satisfied the short-term human interest but not the interests of fellow creatures or even of longer term human concerns.

The same kind of problem occurs in social impacts. If a major project (such as the Trans-Alaska Pipeline) causes high local inflation which then distresses elderly people on low and fixed incomes, it is not enough that the state increases its income by raising oil severance taxes, because that action by itself doesn't compensate the right people. Likewise, fining a polluter doesn't do anything to compensate unidentified victims for real but unmeasured health losses.

In short, preventing unwanted ecologic or social impacts is a far more equitable and usually more efficient strategy for internalizing project costs than transferring money from winners to losers.

The Seal of Society

Any human activity—unless it is part of the persisting shadow side of society—can be sustained only if it meets current social criteria such as justice, equity and morality. In Alaska certain resource practices that meet economic and ecologic tests nevertheless are not tolerated by society at large. The prohibition against the highly efficient salmon traps, whose use in many ways would make management easier, is a well-known instance. In that case, equity was affronted; a few who controlled trap sites could catch nearly all the fish. Codes of ethical conduct such as the rules of good sportsmanship define other arenas where social acceptability, not profitability or environmental soundness, is the main criterion. Often these social codes are written as law or regulation: the requirement that a proportion of annual timber sales in Tongass National Forest be offered only to small companies is an example. Others, such as Alaska's strong inclination to prohibit hiring Outside labor on state-supported resource projects, cannot overcome legal hurdles and must remain less formally codified.

Social acceptability, like profitability and ecologic fitness, is a moving target. Practices acceptable a few years ago are not tolerated now, and some now commonplace may well be illegal in a few years. In the 1950s there was abundant public support for poisoning wolves in Alaska; today, though wolf population reductions are often demanded by some segments of the public, the use of poisons is anathema. Smokey Bear once put out every possible spark; now he carries matches as well as a shovel.

Social acceptability implies that people know about a proposed or ongoing practice, and implies a process of political validation. The decision process itself must be accepted by and the outcomes agreeable to a generalized majority as well as to significantly affected minorities.

In Alaska and other northern places, the site of a resource activity, and the home of the comparatively few people most directly affected, is often physically and culturally far from the place where decisions are made. I have in mind not only the obvious gap between, let's say, a coastal Inupiat village along the Beaufort Sea and Washington, D.C., but the similar gap between

Juneau, Anchorage and Fairbanks, on the one hand, and any rural Alaska community in the path of a proposed resource project. For years the small, non-Native community of Tenakee Springs has fought against logging plans devised in Forest Service offices only 60 airline miles away in Juneau. Natives of Tyonek, just across Cook Inlet from Anchorage, have a bitter running argument with conflicting commercial and sport fishing interests a few miles but social chasms away. The polarities of rural-urban, resident-nonresident, Native and non-Native are ever-present problems in the social validation process for northern resource and land use proposals.

If human settlement of the North is to be sustainable, resilient and life-giving, these social and cultural pluralisms will have to be brought into fuller harmony. Some believe that only when the satellite or "marginal" cultures of the rural North become more like the urban south—when we all wear plastic clothes bought with plastic money—will the problems go away. I suspect that milk is a good deal easier to homogenize than people are, but in any case it is strategically preferable, I think, to protect or even nurture local variation on the human theme, and to find less obliterative ways of mutual problem solving.

An incident revolving around the conservation of wild geese provides an illustration and offers hope.

The fringe of the huge Yukon-Kuskokwim River Delta, inches above the Bering tides and lush with marsh plants in summer, was a fabulous nesting ground for black brant and emperor, white-fronted and Canada geese. Historically, at least 600,000 geese have been raised there annually. Except for the emperor goose, which stays in Alaska, the various species spend winters in California and Mexico, joining others of their kind from other parts of Alaska and western Canada. Populations of all four species plummeted in the 1960s, 1970s and early 1980s; failures on the Delta overwhelmed stable or increasing production elsewhere. Not really knowing the cause of the decline, federal managers nevertheless used the easiest tool they had and reduced legal harvest opportunities, targeting California where by far the most geese had been killed during the past half-century. Numbers of geese continued to drop. The most likely cause—the evidence was circumstantial, but strong—was spring and summer hunting by Yupik people who live in villages scattered throughout the Delta. These hunts were technically in violation of the 1918 Migratory Bird Treaty Act but had been winked at by federal and state officials who from experience knew the political difficulties of enforcement and who knew how ancient, strong and economically important the goose hunting tradition is among the Eskimo people of the region. When the number of residents of villages on the Delta doubled between 1950 and 1980, however, at the same time goose numbers were cut in half, and in half again; and when banning goose hunting in California failed to stem the ebb tide, the harvest by Yupik villages couldn't be ignored.

Among the many steps the U.S. Fish and Wildlife Service took in the 1980s—a massive research effort on the Delta, the hiring of Native technicians and liaison officers, a multipronged public information campaign, and the design of a cooperative management agreement with Yupik leaders—the one I want to highlight is the most interesting in the context of social harmony. In August 1983 the service called and funded a major meeting in Bethel (the large community of the Delta region) to work out the elements of a management strategy. California sport hunters, Delta Yupik, California Department of Fish and Game, Alaska state game managers and the Fish and Wildlife Service all were represented. In November the same groups met at Chevak in the heart of the goose breeding area, and in January 1984 Yupik and other Alaskans met in Sacramento, California while the remnant Yukon-Kuskokwim Delta geese were eating rice shoots in adjacent fields. At that Sacramento meeting the political logjam broke. Later in 1984 an initial agreement was signed, and in 1985 the basic document was expanded and ratified by the cooperators. I can only imagine the melange of responses and experiences the Californians must have had to the tundras around Bethel and Chevak, green and windswept under grey August skies, white with snow in November; or that the Yupik people had in the hot, smoggy, agribusiness landscape of central California. A shock to both, perhaps. But they came together as persons needing to solve a problem, ready to give in order to get, not pressured to deny who they were and what they believed in order to improve the lot of a shared and valued resource.

There are, I suppose, limits to how far the criterion of social acceptability can be applied. Unless we think that contemporary majorities are always able to identify their own best interests, and always will protect vigorously the interests of children to come, we have to assume that societies will occasionally—perhaps often—be wrongheaded. How do we protect ourselves against our own folly? We can put some faith in technical experts to point out where we are going awry, but at the same time we draw back (properly, I think) from giving the technician too much power. As Bernard Devoto once said, experts make fine employees but disastrous masters. We ask, instead, that the technician be persuasive rather than powerful. We also draw on the altruistic bent of society at any one time to protect future interests. Perhaps bureaucratic inertia itself is a partial check on wrongheadedness, as often as it frustrates wisdom. We avoid making too many bad decisions by making every decision so slowly.

Writing to his nephew in November 1787, George Washington expressed a needed humility when he said, "I do not think we are inspired, have more wisdom or possess more virtue than those who will come after us." Nevertheless, we aren't sure we are any less inspired, wise or virtuous than future generations either, and in any case we are all we have. It is just as wrong to do something without social validation as it is to be guided wholly by what now convinces us of its popularity. Social acceptability is a valuable criterion of a resource practice that is, in the foreseeable future, sustainable. Without it, our actions and interactions with nature can only be chaotic.

There is no ultimate best set of individual behaviors or social dogma that will keep people forever in equilibrium with their changing selves and a changeable nature. All we can do is focus on directions that feel right, that seem congruent with what in our time we know of nature, that conserve rare and fragile things, that retain possibility and variety. The guidelines I've summarized here are meant to be fundamentally hopeful, and resilient, not dogmatic prescriptions. They leave a lot to the wisdom of individuals confronted with specific tough choices at particular times and places, and they give us the comfort that, when we make a mistake, there will be fewer and smaller fragments to piece together.

6
Toward Enduring Societies

Firey's criteria for enduring resource practices (that they must be profitable, acceptable and ecologically supportable) and my suggested guidelines for fitting into the northern country are valuable as far as they go. What they need is a fuller context, a broadening and rounding from the narrowness of practices to the realms of societal meaning and decision making. Creating a new flavor of ice cream may be welcomed, profitable and environmentally benign, but is fundamentally trivial. The familiar but chameleonic notion of development is a useful framework for that needed broadening. In its worst form development is the public justification of private greed. To me, development is (or should be) the unfolding process of human betterment that uses instruments of individual, political and economic change. Defined in those admittedly general terms, development provides a hopeful structure onto which appropriate natural resource practices and broader relationships between people and the rest of nature can be built.

Defining Development

Like economists, psychologists and political scientists, biologists have appropriated the term development to describe important phenomena of life. Their use of the word is an illuminating metaphor for our context of social change. Applied to embryonic growth, to postnatal maturation, and to progressive change in communities of interdependent species, biological development has three essential features: a comparatively unformed starting condition with genetic information present, a progressive set of linked changes in which nature and nurture interact, and a fairly predictable end condition that is complex, self-regulating and rather stable until death or a major environmental disturbance occurs.

Embryonic and postnatal development are two phases of the same process, the transition being marked by a great leap in independence of the individual. Controlled by genetic information, the embryo grows predictably from amorphous to differentiated, from unbuffered and relatively nonsensing to self-regulating and sensitive, from absolutely dependent to relatively independent, and—necessarily—from simple to complex. The whole process can be seen as an achievement of potentials present but invisible at the start.

After birth the individual, still strongly influenced by the genome, is continuously reshaped by its experiences in a particular environment. Physically the individual becomes stronger, better coordinated and eventually sexually competent. The developing person moves from reactive to thoughtful, from unaware of self to ego- and other-centered.

Community or ecosystem development is a close analogy. On the new atoll, lava flow, gravel bar or plowed and abandoned field, information is present in the form of surviving species, seeds or early pioneers. New immigrants arrive. The sparsely interacting, almost random collection of plants and animals that begin the sequence slowly becomes more interdependent and has an increasing effect on soil, microclimate and its members' mutually intertwining fates. Biomass and complexity increase. Resources are used more fully. At some point energy and growth are balanced by decay, the loss and addition of species slows, and the whole assemblage—now a true community—has the capacity to maintain its form and structure in the face of normal environmental variation.

Human social development embraces many of the same elements. There is a beginning condition of unachieved potential (not the clean slate Adam and Eve stepped across out of the Garden a million years ago, but any selected starting point in history), a process of progressive change, and—but here we are in trouble. The end points of human development are rarely discussed except by philosophers and theologians. We do not know where, or whether, the process will stabilize, nor what the human condition then will be. We rebel against the very idea of final achievement, preferring the openness of undefined goals and the satisfactions of apparent short-term progress.

Economic development, too often thought of carelessly as an end, is (we hope) a means toward some of those intermediate goals of personal and community development. Economists define development in neatly measurable terms as increases in per capita wealth and in production per unit of utilized land, labor, capital or information. Satisfied with wealth and efficiency as meter sticks, they avoid the questions of whether persons or communities experience an unfolding of potential as a result. Understanding economic development as an instrument, not a goal, is a key to further human progress on spaceship Earth. When we seriously ask Why? we will find new and better answers to How?

For several centuries, including the whole of the Industrial Era to the end of this waning 20th century, Western societies have presumed that the specific and local gains in population, employment, profit and community vigor comprising development were not only permanent, but would usher in further gains. Accumulated bitter experience has shown otherwise. Quite recently, in tacit recognition of the transientness and vulnerability of many such changes, we have begun to specify that development should be sustainable. Writers like Raymond Dasmann[49] gelled the idea two decades ago, and by the late 1970s sustainable development was firmly imbedded in the agenda of the United Nations and all nongovernmental international groups concerned with resources and environment. The World Conservation Strategy (1980) attempted a consensus on the more traditionally "environmental" aspects and enlisted 100 nations in the effort. Lester Brown's "Building a Sustainable Society"[50] (also 1980) gave the concept more comprehensive and practical treatment. In 1987 the World Commission on Environment and Development (the Brundtland Commission) gave sustainable development new political stature in its broad and much-quoted report, "Our Common Future."[51]

In a nutshell, sustainable development does not carry the seeds of its own destruction. More elegantly, sustainable development is, as the Commission wrote, "a process of change in which the exploitation of resources, the direction of investments, the orientation of technological development, and institutional change are all in harmony and enhance both current and future potential to meet human needs and aspirations." Sustainability brings the future into the present by demanding that we try to work out consequences and discharge responsibilities to future generations. If development of one person, community or nation is achieved at an intolerable cost to others, then that change, good as it may seem to beneficiaries, is unstable. If the development is based on extraction of an exhaustible resource and the short-term benefits are frittered away, it is not sustainable. The same is true if the process leads to a poisoning or collapse of critical environmental systems.

There is no fixed set of behaviors or resource uses that we can say is sustainable forever. Market and technological changes, unforeseen ecological effects of human actions, global shifts in climate and changing human values and preferences all can make a once-stable process suddenly unsustainable. All decisions and projects purporting to be sustainable are essentially experimental.

Sustainable Development in the North

The Brundtland Report spans the globe. It establishes a scope of concern and discusses difficulties common to the entire world or major regions within it. The report attempts to define the responsibilities of have and have-not nations equally, though its central interest is in problems and alternatives for the Third World. The Brundtland Commission set the stage for nations and clusters of nations to pick up where the world view left off, tailoring assumptions, issues and priorities to their own land and culturescape. Many nations have begun that process.

Canadians have been especially vigorous in attempting a regional definition of sustainable development, in part because of acute tensions between the settled South and remote North, partly because of long-standing unease in the South over the extent and meaning of U.S. and other foreign investments in Canada. Thoroughly involved in the creation of the World Conservation Strategy, Canada hosted an international assessment of that effort in 1986. By 1987 every province and territory had articulated its initial views on conservation and sustainable development—the two by then being scarcely separable. Specifically northern issues and strategies were addressed in a conference in Vancouver, B.C. in 1988. At an even finer level, by 1988 there were reports on development strategies for individual settlements (Old Crow, Yukon, is an example), for Inuvialuit people in arctic Canada, and for physiographic regions such as the Canadian Polar Basin.

For all their diversity these documents have common themes that blend and interrelate. Most are in the voice of the North, speaking to the South. The North, they plead, is physically, biologically and culturally different. These differences, pervasive and important, are not yet well enough understood or incorporated into industrial or governmental planning. Northerners feel like colonials, controlled and patronized by southern institutions. Unanimously, they ask for heightened self-determination. Indigenous people feel especially misunderstood, ignored and trampled. What makes the North a colony in feel if not in political fact is the power of economic domination in the form of huge energy and mineral extraction projects. Northerners assert that they have little chance to negotiate critical details of those projects (when, where, how) and no power to say whether they should go forward at all. People of the North tend to place priority on renewable resources, while the South is almost entirely interested in nonrenewable ones in the North (an exception being hydroelectric power). And finally, residents of Canada's North insist that if bureaucracies are fragmented while real problems of resource use, land care, culture integrity and community well-being are all-of-a-piece, it is up to the institutions to change.

These themes describe, for northern Canadians, what sustainable development is not. It is not remotely initiated and controlled, culturally dissonant, environmentally destructive, unresponsive or shortsighted. What attempts have been made at positive goals and action recommendations tend to be the flip side of the criticisms. For example, local residents often demand, and sometimes have gotten, special cooperative structures combining local and regional or national interests to resolve an issue once in the domain of a distant agency.

The Brundtland Commission doesn't confront common world views in any profound way. It is far too comfortably utilitarian and human-centered for Deep Ecologists, who simply say there are too many people and too few of other creatures. Its tone is sober, even grave, but it implies that with serious effort people of reason and good will can work everything out. It has an air of common sense, and even when its criticisms of present ways of doing things seem daring, sustainable development is ringing bells that have been rung before, albeit less prestigiously. Too often it seems to accept the old and destructive premise that growth is necessary for development—which creates the oxymoronic phrase "sustainable growth."

None of this is meant to denigrate the idea or the cause. Its challenges for change put it far enough ahead of *realpolitik* as to be almost revolutionary. Certainly the work it defines is both worthy and enormous. Those who have been the most articulate sketchers of sustainable development as a concept have been careful to preserve broad options of interpretation, priorities and tools for people of each nation. Perhaps most important, sustainable development insists that people keep problems in the round, not artificially fragmented into deceptively isolated areas such as land abuse, industrial investment, poverty, environmental protection, overpopulation and urbanization.

The Idea of Carrying Capacity

"Our Common Future" and many of the national and regional reports it has so far catalyzed shy from a term commonly applied to global resource and environmental problems in the 1970s: carrying capacity, the notion that at some level of human abundance we will bump against an ultimate ceiling. Perhaps carrying capacity, like so many words in political debate, already has been crippled by glib and careless use. Maybe it sounds negative and rigid, and spawns frustration instead of action. Or—and this is a third possibility—people have come to doubt whether it should be applied at all to humans. Nevertheless, the idea of land and resource limits persists. It is intuitively commonsensical, and many everyday experiences point to its truth. Local residents often have a very immediate sense of population limits, at least with respect to visitors and immigrants. Throughout the far North people in rural settlements fear incursions of oil field or mine workers or of sportsmen, who they think will unbearably compete with residents for scarce fish, game, firewood or other land resources. Not a small handful of rural communities in the Alaska Bush are tacitly understood to be closed to outsiders.

The 18th century hadn't quite ended when Malthus made his famous—or infamous—statement that human numbers, increasing faster than food supplies, would meet an irresistible barrier marked by poverty, pestilence and starvation. Malthus's ghost still paces the market in Guildford, England, in silent frustration. The 1950s and 1960s saw a sharp upswing in neo-Malthusian predictions, but the Green Revolution (or the increased food production with which that program was credited) took the wind out of their sails. Whether cultural evolutionary processes have freed *Homo sapiens* from the biophysical constraints operating on cows in pastures and fish in lakes is still hotly debated. Many human societies (Haiti is an example) are now overtaking their current resource base and paying the price in social disintegration, poverty, starvation and depopulation. At another level, overarching ecological conditions still set broad (arguably distant) limits to the settlement of Earth. On the other hand, local resource availability explains almost nothing about the size or location of any of the world's cities, and global averages of food supply per capita recently rose for five years even while our population passed the five billion mark.

Unquestionably, the ecologist's concept of carrying capacity can't be applied unchanged to the human condition. Humans differ too much from other animals. For one thing other animals exploit environments with genetically programmed equipment that changes immeasurably from one century to the next. As Reynard the Fox was known to the storytellers of medieval Europe, so is Reynard today, still unable to fly to the higher grapes. In contrast, human exploitive equipment changes with lightning speed. Within the horizon of our own impatience, the gestation period while necessity mothers invention may seem long, but in evolutionary terms no time at all elapses.

Second, animals must get everything they need from the area that individuals can reach. This can be very tiny, as with barnacles, or very large, as with the resource supply points of a migratory bird, strung like pearls on an unclasped necklace, spanning thousands of miles. In both cases the genetically defined limits of individual capability set the boundaries. Through cooperative trade, humans transcend such limits. Our settlements can occur in places where only a fraction of needed resources can be obtained locally. In fact, this is the normal—even universal—condition of modern cities.

Another difference is that resource consumption rates don't vary much among mature individuals in other species, but among people consumption varies enormously. This is true not only of calories eaten, but even more of nonfood items like energy, metals, shelter, packaging, vehicles, and so on. Obviously, the kinds of resources demanded, as well as amounts, differ among people with different life-styles and means. Not only that, but resource demands change quickly over time: new plant-derived medicines, new foodstuffs, new fibers, metals and manufactured chemicals— the list is long. Industrial cultures use more elements of a given landscape and of the whole globe than hunter-gatherer or farm cultures ever did, or would. From time to time resources are exhausted or become obsolete, while invention, discovery and price changes create others. Thus, humanly defined resources expand and contract, and with them carrying capacity.

Finally, environmental variation is the only factor significantly changing an area's carrying capacity for animals other than people. With our storage mechanisms (including money and credit) and ability to substitute one resource for another, environmental change is not always important in the short term. In the longer run, however, global pollution, global and regional climate changes, and changing sea levels may be powerful forces in setting new limits to the distribution and abundance of people.

Despite these dramatic and basic differences in resource relationships between humans and other animals, I think that carrying capacity is still a valuable idea. We can only use and trade that which is already on Earth, a planet of a certain volume and composition. We all breathe one atmosphere (as northern Alaska Inupiat inhaling the thinned and transported fumes of smokestacks in northeastern Europe so keenly appreciate). Even on a fine scale of geography, the idea of limits is important. In any local planning forum, we may realize that within the time horizon of

practical decision making, human resource demands may over-
shoot supplies. Something has to change. We can reduce demand,
expand exploitive technologies or find resource substitutes, establishing a new and temporary relation between human community
and natural resources. Understood that way, carrying capacity is a
test of the current sustainability of human behaviors in a specific
human habitat. We know that we are mining out our arable soil and
washing it into the ocean. Agricultural production exceeds ecosystem capacity, not inevitably or immutably, but under the particular
set of demands, tools and policies now predominating.

In the words of "Our Common Future,"

> Growth has no set limits in terms of population or
> resource use beyond which lies ecological disaster.
> Different limits hold for the use of energy, materi-
> als, water, and land...The accumulation of knowl-
> edge and development of technology can enhance
> the carrying capacity of the resource base. But ulti-
> mate limits there are, and sustainability demands
> that long before these are reached, the world must
> ensure equitable access to the constrained resource
> and reorient technological efforts to relieve the prob-
> lem. (p. 45).

The idea that wants—especially in the First World—might be
reduced to "relieve the problem" is conspicuously absent. However, the Commission's report does highlight several key points:
ultimate limits do exist; carrying capacity is, within bounds, mal-
leable; and early action to draw back from the brink of collapse is
a crucial strategy of survival. Ecologists know that a comparatively
small group of animal species often limit their numbers, through
self-regulation, to a level below physical resource maxima. Most
species, in contrast, are controlled by external forces and limits.
Which group will humans join? On a global scale, only recently
have we had both the necessity and opportunity to choose.

Alaska: A More Urban North

All of the conference talk, commission work, reports, government documents and earnest persuasion in international circles,

associated with the popularizing of sustainable development, seem to have left no more impression on Alaskan life than a breeze passing over a fell field. Even neighboring northern Canada's activity has not triggered any response from Alaska's political leaders—though people in many Alaskan villages are participants in the effort to redefine northern issues and raise their priority in political agendas. To be fair, it may be true that urban southern Canada is equally as uninterested and untouched as the majority of Alaska. Only the vastly greater geographic extent and more extreme remoteness of Canada's North has forced sustainable development into the consciousness of regional leaders and specialized federal agencies.

Alaska also may be a no-man's land for the present targets of sustainable development. We are too modernly urban to be in the Third World, too small and sparsely settled to be responsive to the call for new institutional designs and new investment policies aimed at central governments and massive corporations of the industrial world. None of this means that sustainable development itself is inappropriate for Alaska: far from it. It merely means that we need a home-grown version of it.

It will take far more and better minds than mine, and more lifetimes than I have, to create a comprehensive strategy for durable development in Alaska. The thoughts I offer here, and the guidelines suggested in the previous chapter, incomplete and preliminary as they are, may still illustrate directions and issues in a useful way. They all reflect an attempt to weave essential characteristics of nature in the North into the tapestry of social action and public policy.

Bridges in Rough Terrain

One of the most common collective hopes of members of an up-and-down economy like Alaska's is that somehow economic peaks and valleys can be smoothed out. Popular preference, not surprisingly, is to smooth out at a high level, certainly no lower than the current or best-remembered peak. Mistily envisioned, the state's economy would be a series of high plateaus separated by steps of growth to the next sustained level. New petroleum and mineral reserves would be discovered to replace exhausted ones. Renewable resources would be used more thoroughly and in increasingly diverse ways.

Reality, I think, suggests that Alaska's curve of economic activity will mimic the silhouette of the Alaskan Range against a winter sun, rather than a stairway to Utopia. I have already remarked why the economic terrain ahead still looks rugged. It is inherent in northern environments to vary widely over time, and inherent in species in Alaskan ecosystems to change from abundance to scarcity in response. It is a near-certainty that Alaska will always be a price taker—that is, we will sell resources at prices set in world trade—and a costly place to do business, hence vulnerable to market contractions.

One strategy in a northern survival kit, therefore, is the ability to bridge across the canyons. This has to be done by both collective and individual action.

One public bridge, the Alaska Permanent Fund, is already in place. This Fund is fed by an automatic tithe of about 12 percent of mineral (in practical terms, this today means oil) revenues, by occasional special deposits by the Legislature when surplus funds are available, and by uncommitted Fund earnings. In its first decade of existence (to 1986), the Permanent Fund received 22 percent of all state oil revenues. The principal cannot be used unless the state constitution is first amended. Investment earnings are used for inflation-proofing the principal and for any further use the Legislature decides, including (from 1983 at least through 1991) a payment of annual dividends to every individual Alaskan. The principal had risen by 1991 to almost $12 billion, bigger than any other endowment fund or private foundation in the United States. The popularity of the Fund as a rainy-day savings account and as a yielder of annual cash dividends has kept the Fund out of reach of prospective raids during the current economic slump in Alaska.

Clearly an important hedge against future needs, the Permanent Fund is just as obviously not a panacea. Its earnings—currently just over $1 billion—barely meet the demands of inflation-proofing and the $500 million dividend program, let alone a significant fraction of state operating and capital expenditures which are now on the order of $2 billion annually. In 1989 inflation-proofing soaked up 41 percent of Fund earnings, and the dividend program,

53 percent, leaving little for spending or reinvestment. Moreover, the very factors that can force us to dip into the account (declines in oil prices and their rippling economic effects) also reduce income into the Fund. In 1985-87 the loss of oil revenues was offset by exceptionally high rates of return on investments of principal. We won't always be so lucky. And I doubt that our Fund investments would miraculously escape the effects of a widespread national or international depression. Beyond these concerns, it is also true that we have no tested strategies for using the Fund as a bridge across depression. Must we always keep the principal sacrosanct? When is it wise to draw the principal down—if ever? Should there be limits to the rate and depth of drawdown? Where should the money go, to be the greatest possible help at least cost? Should any drawdown at all be permitted while (as has been true beginning in 1981) the state levies no personal income, property or sales taxes? Should the dividend program be scrapped, letting 60 legislators and a governor, instead of a half-million Alaskans individually, decide how to spend that half-billion dollars?

It is easy to forget that the material for this economic bridge comes from stripping the top from peaks. People forego some present income or services during boom times to lay away savings. There is always an ideological debate over whether collective savings by state action is preferable to private savings. In fact, there are and should be both. Because not everyone benefits, certainly not equally, during boom times, public savings can be a way of redistributing the saved revenue according to later individual need. Given Alaska's high transient population of investors, entrepreneurs and workers, public savings may also assure that a lower proportion of good-times profits go South. On the other hand, saving and later spending money in the public sector inevitably incurs losses in the friction of bureaucratic process. Finally, the incentive for common sense and long-range vision in private wealth management ought not be completely dulled by state patronage.

Three Canadians recently studied the Alaska Permanent Fund's strategy of national-international investment portfolio management and found that it has been much more successful than the local investment programs of Alaska native corporations handling money during the same period from land claims settlements.

Rather than investing capital in northern business ventures that in themselves are inherently risky, there is wisdom in accumulating capital until a stream of interest earnings is available. Those can be channelled into small-scale, locally controlled ventures carefully selected to offer a good chance of profitability and sustainability, and to meet social criteria of small northern communities.[52] Surely some would fail, but the failure would not touch the trust fund principal.

One of the flaws in the Alaska Permanent Fund process is that Alaska has been so eager for mining to occur that it has leased mineral lands and encouraged mining at times when profitability is—at best—low. The state's royalty income from nonpetroleum mining is almost invisibly small. When, in 1988, a state agency proposed to raise coal royalties to the legally required fair-market level, a political furor resulted in legislative intervention and the acceptance of a far smaller increase. And in 1989 the Legislature, forced to recognize that miners *leased* rights to mine on (but did not own title to) state lands and must be assessed a royalty, opted for a minimalist royalty formula. In essence contemporary politics favors private profit over public revenue in the coal and metallic minerals industries, perhaps under the assumption that even if the state gives away its minerals, mines employ people and circulate money in the private sector. The trade-off, particularly in the absence of state income taxes, is that the minerals are lost from the state's portfolio of assets without significant revenue for public services or savings. If the state acted more like a profit-maximizing resource owner, there would in the long run be more material for economic bridge-building.

There have been opportunities for local governments to develop investment funds functionally similar to the Permanent Fund—but few have. The chance usually has come early in a major spurt of local growth based on a big new industrial project. When the size of the community is comparatively small and major new sources of property tax revenues are being built, enough money may come in to allow a provident local leadership to channel some out of current spending and into long-term savings. Later it may be too late: local population growth and political commitments to major expansions in public buildings and services can quickly soak up any theoretical surplus. I can think of four recent Alaskan examples. When the

Kenai Borough first formed, it taxed numerous local oil field and refining facilities; these could have yielded savable revenues, but all were spent. The North Slope Borough was formed primarily because of the chance to levy property taxes on the huge and expanding Prudhoe Bay facilities. By 1989 the Borough had a general fund surplus of over $700 million (for a population of under 20,000 residents, that's not bad!). The Borough also had a bonded indebtedness of close to $900 million, stemming from bonds sold to finance public facilities, which meant substantial annual interest payments. A third opportunity came to Dutch Harbor on Unalaska Island, which saw an amazing boom in fish processing and vessel supply activities in the early 1980s. Finally, the brand new Northwest Arctic Borough drew its boundaries to include a huge mineral prospect that, when in full operation early in the 1990s, will become the world's biggest zinc mine. This, too, could yield enough property tax revenues (or, since the Borough is a partner in the mine, profit revenues) to support a savings program.

There are many tax gimmicks, loan programs and savings-bond ideas the state could use to encourage private savings to smooth out cyclic economies. But there is, as well, a totally different approach available: nurturing a generalist or multiple-skill approach to personal employment preparation. A woman or man who can do two or three things quite well has a kind of insurance policy for declines in local or sector-specific economies that a one-skill person does not. This is a good idea anywhere, but especially so in the North. I know a good many people using this strategy who have found economic security over three decades of good times and bad. · Communities could play a much more active role than they do now in fostering this lifeway. For example, a 2-year roving apprentice program, publicly financed, could give young high school graduates practical experience in a half-dozen or so jobs.

Foraging, Wage Work and Creative Unemployment

In our public consciousness we carefully measure and coddle one limb, the commercial corporate sector, while we ignore and erode another. That other limb of our economy, the informal economy of voluntary, cooperative, barter-based and household activities, can be very large and rich but is essentially transparent to the minions of GNP and the Dow-Jones index. In their fine but little-known book "From the Roots Up: Economic Development as if Community Mattered,"[53] David Ross and Peter Usher estimate that in Canada the informal economy is at least half the size of officially recorded GNP.

This ghost economy enfolds three main sectors. One is the critical, universal, unpaid work of household maintenance, often involving what economists (if they were interested) would call import substitutions like gardening and the making of clothes, tools and toys. Another is creative unemployment, the hours or months or even years when a person's unpaid efforts in art, invention, community service or self-improvement may yield the greatest of contributions to human development. Last, and particularly in Alaska and northern Canada, the informal economy includes the foraging or subsistence sector.

The oldest and largest aspect of the foraging economy in the North is the traditional economy of rural Native villages in which families, in season, gather food, fuel, and shelter- and tool-making materials from seas, rivers and uplands. It is a partly ancient, partly modern activity. The gathering, making and consuming often are accomplished with the help and companionship of old stories, songs and rituals. Tasks are done according to ancient teachings of right and wrong, blending the pragmatic and sacral. They are also aided by rifles, skiffs, motors and snowmobiles—sometimes even aircraft—bought by earnings from temporary wage employment and from transfer payments from governments. It is this village economy that in Alaska and Canada has received so much belated attention in recent political encounters over land claims, land use and Native sovereignty.

A different dimension of the foraging economy, the neotraditional aspect, has been entirely ignored by economists and politicians alike. Throughout Alaska, in city, town and countryside, people directly gather the yield of the land. A family in Fairbanks picks five gallons of blueberries on a nearby hilltop in a weekend afternoon. A boy in Homer digs a bucket of clams during a September ebb tide. In Juneau two men buck up a cord of firewood from spruce trees recently uprooted by wind along a commuter's road. A Nome storekeeper leaves town early on an October Saturday for one more whitefish netting expedition; by evening his smokehouse is full of savory fillets. From everywhere, to almost everyone, the country gives its bounty of mussels and mushrooms, birch logs and bear meat, crabs and caribou. Whether the casual fun of an urban family or the adopted life-style of a Pittsburgh expatriate, the neotraditional foraging economy pervades northern society. According to scientists in the Alaska Department of Fish and Game, the median harvest of wild food per Alaskan is 250 pounds per year. Some rural Alaskans harvest more fish and wild game than the average American buys each year in a grocery store,[54] not because they have gargantuan appetites but because they are foraging for a family.

Officialdom has been painfully reluctant to admit the existence of the foraging economy. Generations of northern bureaucrats have assumed that the subsistence economy is anachronistic, important only to a remnant population beyond the frontier. According to this political wisdom, foraging inevitably and quite properly will disappear into the maw of the industrial economy. The task of the benevolent administrator has been to encourage the transition of subsistence practitioners into the mainstream economy: hence vocational schools, a strongly biased public education curriculum, and so on. That the transition, now close to 150 years old, should take so long has caused little wonder and no enlightenment. Only in the past 10 years have enough scholars reported on the foraging economy as it actually exists, not as it is supposed by urban minds to occur, to allow a more accurate picture of its extent and meaning. We're beginning to realize how durable and adaptable the subsistence life-style is in response to the massive resource projects that were supposed to be the vehicles carrying indigenous people into modern life. To be sure, it is not the pure foraging economy of the Stone Age. But today's mixed cash-subsistence economy is as far from the market economy of Anchorage, Edmonton and Des Moines as it is from its own early roots. It is a hybrid of surprising vigor.

The modern hybrid subsistence economy is now, and I feel strongly should remain, a rewarding way of living that people should be able to choose. I don't imply that there is infinite room for more participants, any more than there is for farmers or computer technicians. Ultimately limited by natural productivity, foraging niches are often immediately constrained by custom, competition and cost. As long as such a lifeway exists, however, some can adopt or inherit it even as former participants die or move out of it: a kind of informal limited entry system.

The foraging economy deserves recognition, respect and collective nurturing. Its fundamental and radically different precepts of home production, wide sharing and conflict avoidance by their very existence enrich the culture of a commerce-based nation. It may have strategic importance as a kind of lifeboat for the skilled few who within it can weather the falling barometers that typify far northern industrial economies. Then, too, sharers in the subsistence process are a potent constituency for good land stewardship, a salt-of-the-earth constituency often less easily ignored than the "elite city folk" who traditionally lobby for conservation.

The primary requirements for nurturing the foraging economy are easily sketched in general form, however mischievous or stubborn the problems of implementation may be.

The clearest call is to care for the living landscapes and seas from which the harvests are gathered. There is nothing mysterious about what needs to be done. In the wild settings of most subsistence activities, the best and simplest stewardship strategy would be to leave the land alone. To put this ideal of benign nonmanagement into practice presupposes a political decision to designate vast areas as inviolate subsistence lifeway reserves—an act that likely would be opposed by many, including some subsistence resource harvesters themselves. Anything short of that plunges us into the complexities of multiple use management and the erosive forces of repeated reasonable compromise. As far as I know, no northern highway, mine, pipeline, oil field, dam or urban expansion project ever has increased the subsistence resource base; they differ only in how much they diminish it. There is every reason to try to make each authorized intrusion less destructive, and over the longer run net losses may reach zero as the effects of new projects are balanced by the reclamation and healing of old ones.

To make a least-impact approach as successful as possible, the essential prerequisite is to know what can be lost least painfully and what must be protected most staunchly. Up to now Alaskans have mainly relied on hard-won strokes of the broadest political brushes to sketch programs of earthcare. We name huge areas as refuges and parks, hoping that the critical ecosystems are within their gerrymandered boundaries. Statutes setting permissible maxima for water and air pollution, establishing criteria for forest management, and so on, are cut from the same cloth. They do not prevent a locally tailored approach, but neither their language nor the budgets offered to carry them out encourage it. Northern science is able now to be a helpful partner in the priority-setting process, and the extraordinary knowledge of local hunters, fishers, and gatherers is equally potent. We know that ice-front environments in the Bering Sea, wetlands of floodplains, tundra and boreal streamsides, marshy coastal estuaries, streams hosting large salmon runs, and warm-soil slopes of the hilly Interior are places of high energy and nutrient transfer. This detailed sense of local priorities needs to be brought into public policy as well as stewardship practice.

The soils, waters, plants and animals that are the *sine qua non* of the foraging economy require permanent and locally implemented protection. The next obvious need is to assure that they are dependably available to people within the subsistence sphere. It is neither likely nor, I think, wise to strive to set up an all-encompassing, exclusive subsistence right of access to those resources. To do so would require eliminating all competitors: the commercial fishermen, sawmills, recreational fishers and hunters, perhaps even touring bird-watchers and backpackers, who have their own worth, validity and customary rights under the system of governance that now exists. What is both essential and possible is to agree on some broad fair share for foragers: "no less than we now have" would be a minimalist solution. The renewable resources now used by the subsistence economy already are only the remnant after a century of almost unrestrained commitments to the needs and wants of the industrial society.

Since 1978 Alaskan and federal legislators have tried to establish some level of legal certainty for subsistence. Federal actions have permitted subsistence harvests of otherwise-protected marine mammals and moved toward legalizing traditional summer goose and duck harvests. Federal law requires federal landholding agencies to consider effects that proposed land uses would have on subsistence resources. National statute also sets forth elements of subsistence resource management Alaska must embody in state law to prevent a take-over of management on federal lands. The state has responded with a law and regulatory regime which, on its face, gives subsistence users first chance at wildlife harvests. People meeting criteria such as rural residency and historical and customary dependency on wildlife legally become subsistence users who are given priority in the harvest of animals identified as subsistence resources. Commercial and recreational harvests are allowed when officially prescribed subsistence needs are met. Rural people are formally involved in annual regulation-setting. (The criterion of rural residency recently was found by a state court to be unconstitutionally discriminatory, and in July 1990 the federal government officially took over authority for subsistence management on federal lands.)

The system is better than none, but it has serious flaws. One is the incredible complexity of the regulatory structure. Basically stemming from the impossibility of writing static definitions and prescriptions for a diffuse and ever-changing human behavior, the subsistence regime at every turn shows the struggle of lawyers to win lawsuits, politicians to prevent clear blame, and biologist-managers to permeate the whole with the rationality of science. The result, so far, is an endless bickering over definitions and political intent and a frenetic search for biological and social data at ever finer scales of time and space. The other basic flaw is that neither this regime nor any or all other federal or state laws combined set any firm minimum size of the foraging resource. There is little advantage to sitting at the head of an empty banquet table.

The family is the central unit of both production and consumption in the traditional subsistence economy. Even in the hybrid modern version, the family is still central, perhaps because much of the wage work is aimed at earning cash for foraging equipment and the hardware of food preservation, storage and preparation. The subsistence family, unlike the industrial family, is economically as well as socially fundamental. A broad range of social issues needs to be reviewed in this light. It may have been an even worse mistake than we now think it was in the decades after white settlement of Alaska to force village children to leave home communities for distant schools. The 1976 judicial and later follow-up legislative decisions to build more high schools in rural Alaska, even at a high and perhaps unsustainable cost, unwittingly may have supported the foraging lifeway. Drug and alcohol addiction prevention, amply justified on other grounds, likewise are congruent with family

integrity and successful subsistence. Unfortunately, many programs and forces are disintegrative in effect. Even the subsistence
regulatory regime may be treacherous, based as it is on individual
licensing and individual harvest limits rather than on real social
sharing-groups.

I spoke earlier of the need to grant the subsistence lifeway full
and merited respect. This may well be the most basic and difficult
change our society must make. We persist in thinking of subsistence as primitive. We berate it for being impure in its modern form,
when in fact it is highly adaptive. We think of it variously as a hold-
over, as economically inefficient, as a sink for subsidies, as the last
resort for the unskilled or lazy. As a land use we seem to tolerate it
only as long as there is no competing demand like oil extraction,
mining or commercial timbering. All these follow logically from
disrespect. All need to be reversed if an enriching, unique and
strategically important lifeway is to be an enduring part of northern
futures.

When I moved to Alaska over three decades ago and began
absorbing the thought and action patterns of people here, one of the
oft-repeated complaints about the local economy was that most
jobs were seasonal. Construction, fishing, logging, farming, fire
fighting, tourist serving: all have their months of activity and their
doldrums. To community leaders and economists, this not only
seems bad from a standpoint of overall economic volumes, but an
affront to the doctrine that only consistent hard work earns success
and salvation. To eliminate seasonality has long been part of urban
Alaska's economic agenda. Rural places, more closely bonded with
natural calendars, seemed much less excited about the problem. In
fact—and many encounters over the years confirm this—many of
those with seasonal jobs prefer life that way. While there are
disadvantages to on-off employment (the chief one being that you
tend not to amass too much property), there are individual and
social aspects of seasonality that are attractive.

For one thing daily earnings tend to be high. Bosses need to cram
so much production or sales volume into such a short time that they
employ help for long hours, sometimes at premium wages. In
many isolated worksites savings can be high, if the poker games
can be avoided. The seasonal savings, of course (along with unemployment compensation), have to support life in the intervals
between jobs.

Mainly, however, there seem to be two clear benefits to a northern society in valuing and supporting seasonal work. One is that people who work at seasonal jobs often develop two or three salable skills, hence becoming generalists to a degree. Their individual flexibility in employment, as well as the survival skills they learn during times of unemployment, are part of the resilience northern societies need. The other is that unemployment itself can provide the chance for creative contributions to society. Volunteer community service is possible for sustained and substantial blocks of time. A woman I know earns enough money with seasonal carpentering to support her volunteer Witness for Peace work in Central America. Many writers work for wages just long enough to support their cherished times of intense focus on poems and stories. The same is true of numerous potters, painters, weavers and other artists and artisans. Especially in communities too small to support paid positions of public service, society benefits immensely from the volunteers whom the Bureau of Labor Statistics puts on the red side of the ledger, but who are in reality very creatively employed.

The Bioregion

In the never-ending human game of "Wouldn't It Be Better If...," a recurring proposition is bioregionalism: the notion that areas of similar nature should be central organizing spatial themes in human affairs. At one extreme bioregionalism is merely an expanded boundary for specific kinds of multi-state planning, such as river basin projects. At the other it is the fervent but wistful utopia where people throw off the burden of conventional political boundaries and reaffirm their spiritual, organic and economic ties to a home land.

It doesn't take long with an atlas on your knee to see that natural unities historically have shaped human politics at every scale. A bioregionalist redrawing the world map might very well reinvent the nations of Japan, India, Australia, Chile and Great Britain, for instance, or, within an arbitrarily defined Canada, identify environmentally logical provinces of Newfoundland and British Columbia. On the other hand, political determinism unarguably has prevailed in the layout of most nations of the world, especially Europe, the Middle East, Africa and Central America.

Bioregionalism has been criticized on two main grounds. First, it is impractical, its critics say. It is too late to change political boundaries; the down-and-dirty politics of entrenched interests will never allow their accustomed alliances and adversaries to be reshuffled, lest they lose power. A good many hopeful starts on regional economic and river management schemes have been blunted and dissipated in exactly that way.[55] Second, critics charge that bioregions are doomed to failure by their one-sided view of what makes societies tick. Yet, environmental determinism—even if a fair description of what I understand to be a broad, many-fibered fabric—surely can be no more blind than political or economic determinism. The bioregion based on unities of climate, physiography, and natural resource complexes would be much more durable as a scale for decision and action than the wandering sands of politics or the fickle partnerships of commerce.

My particular interest in bioregionalism is in exploring how its ideas might enhance the adaptive fitness between nature and society that seems so crucial to lasting human development in Alaska. We need, I think, a level of decision making and integration between the local and the statewide. At this middle level the generalized needs of the state (and nation, in its public land resource management and environmental protection functions) would be made more specific by the collective decisions and advice of communities sharing life in a single realm of nature. At this level, too, communities would find the strength to ward off common threats and achieve common hopes.

The closest we have come so far toward a regional level of integration is to implement the 1971 Alaska Native Claims Settlement Act through 12 Native regional corporations that together encompass the whole state. These regions were defined to fit the distribution of ethnic groups of Indian, Aleut and Eskimo people. Because these ethnic groupings arose largely from the slow creation of lifeways fitting the demands and potentials of broadly different landscapes, the 12 regions express ecology and culture almost equally. The act identified villages within each region and defined the village-region relationship. It established business corporations at the regional level and let villages choose whether to establish profit or nonprofit corporations. Given that every Native Alaskan is a stockholder in a regional corporation, and that in many rural regions a large majority of residents are Native, the power of the regional corporations to find and act on the collective regional will is obvious. These Native regions, however, are not models for regional governance, as fitting as their boundaries may be from a bioregional viewpoint. They do not represent non-Native interests at all (and 85 percent of Alaskans are non-Native), and the regional for-profit corporations only represent the commercial interests of their shareholders, not their social or even general economic interests.

The borough is the official level of government in Alaska between the single community and the state. Similar to counties in most other United States, boroughs have the power to tax, receive state revenue sharing and other grants, and (if incorporated with the specific power to do so) plan and zone land uses. They provide school, regulatory, local road maintenance and other common services. Several Alaskan boroughs have large landholdings which they selected from state lands within their boundaries under a legislatively set formula. Alaskan boroughs range from small ones that are almost entirely urban, such as the Anchorage and Juneau city-borough governments, to those that spread over thousands of square miles of hinterland, such as the North Slope Borough.

In contrast to counties, boroughs and parishes in other states, boroughs do not blanket Alaska. In fact, in 1991 only 35 percent of Alaska's geography was (but at least 85 percent of its citizens were) within a borough. The rest of the state is unincorporated, a kind of amorphous pie from which slices may be taken on local initiative (with state approval) to form a new borough. Rural Alaskans are quite suspicious of government and are not exactly eager to create what might become a bureaucratic Frankenstein. The two newest boroughs, both very large, were created when a rich potential property tax base arose in the form of the Prudhoe oil fields and the enormous silver-lead-zinc Red Dog mine under development just inland from the Chukchi Sea.

Whether it is already too late to reshape boroughs along bioregional lines is debatable, but in the long view I think not. At least three current boroughs—North Slope, Northwest Arctic, and Kodiak—are congruent with Settlement Act regions and with logical bioregional boundaries. Interestingly, two of four standards for borough incorporation have a distinctly bioregional tone:

one requires that "the population of the area is interrelated and integrated as to its social, cultural, and economic activities...," the other that the boundaries "conform to natural geography." Late in 1989 the state decided to begin a project to find out how Alaskans want to draw boundaries of boroughs in the present unincorporated lands. It will be interesting to see whether bioregional ideas are expressed successfully.

The important riddle of boundaries, however, is not where they should be, but what they are meant to hold in or keep out. For the regional planner the bioregion simply brings two badly aligned binocular tubes back into focus so that the human community of concern matches the natural resource (hydroelectric potential of a watershed, for instance) it is concerned about. If it comes to a vote, the whole votes on the basis of its full scope of interest rather than each slice on its own isolated piece. The regional planner's boundaries imply no substantial change in societies or economics and make no pretext of usurping the importance of traditional political divisions.

The utopian bioregionalist is a revolutionary.[56] Bioregionalism in this context is a drastically different way of life in which people rely mainly on resources available in the defined region, emphasize personal, community and regional self-reliance, favor the devolution of power from big governments and corporations to smaller ones, emphasize cultural and natural diversity, and work within a steady-state economy. The bioregion, to them, is the main center for decisions about how resources are used, by whom and for what purposes. (Note how innocuous this sounds, but how different from contemporary Anywhere, whether Alaska or Brazil, and how hard it is to imagine, immersed as we are in the colonialism of distant power!) The boundaries of that kind of a bioregion, it seems, would somehow have to be stronger if only to allow self-reliance and self-determination a territory in which to work.

Can the bioregional person, striving for self-reliance, escape being self-centered? We know more surely than ever that the individual affairs of over five billion humans are having effects on Gaea itself, not by mere addition, but also by the multipliers of synergy. Are big structures of study, education, planning, decision making and even coercion on a global scale a practical necessity? Those are questions we have to face with or without bioregionalism, of course, but the modern Westerner is already used to living in the shadow of bigness, where the postmodern bioregionalist would rebel.

Bioregionalism balances nature-centeredness and culture-centeredness. Fundamentally, a good bioregion readily attracts and serves the emotional, spiritual, economic and social needs of the people within it. It is an important scale of personal attachment and pride, of economic sustenance and of collective history. It is a place where a person can still make a difference, and the difference is measurably worth making. The bioregion is an area where the tied-togetherness of natural systems provides a foundation for stewardship of land. In turn, the unity of natural expressions within the region nudges human affairs toward the same unity. An optimal region will centralize those decisions that, taken community by independent community, are too parochial, friction-laden and isolative. At the same time it will accept responsibilities made now at state and national levels. An optimal region will become a unit of human adaptation and dynamic cultural and economic diversity. In a worthwhile region, poetry and produce alike will flow out of the specialness of the place and evolve mutually with it. The place, the bioregion, will be the space in which dwellers will feel at home, *inside*, and confidently grounded.

Ways of Knowing

An important attribute of a bioregion is that it can be known. Its important features and character are accessible through direct experience: not the hop-scotch experience of quick visits, but the intense and rich experiences of residence. Furthermore, the region is large enough to have a history worth telling, a history that is worth listening to. It is the largest geographic area in whose evolution and development the ordinary resident can participate out of personal knowledge. A state as big as Alaska, and certainly a nation as big as the U.S., can only be known as a whole by a tiny minority after a long life of extraordinary opportunity to travel and observe. Even then the knowledge is—must be—dangerously shallow. Such big areas can be only perceived statistically: so many square miles, so many copper mines, such and such a national debt.

In passing—though the point is not trivial—I think the idea of participating out of knowledge is worth a second look. If the region is to meet its potential as a significant level of social decision making and relating to the land, it is critical, I think, that the wide variety of ways of knowing be recognized and used in the creation of regional development strategies as well as in pursuing them through action. As Barry Lopez commented in "Arctic Dreams,"[57] the countryside is not, it is perceived as. Reality, which is always much more than is being recorded about it, is not only what is being sensed, but who is sensing it. Inherent in the "who" is a set of purposes, which determine means, which determine the facts selectively recorded.

Myths and stories are a means of knowing. By allegory and metaphor they distill essences from generations of experience and adaptation, and teach not only a cultural and moral framework for our own experience but also specific ways of getting along in the part of the world to which the myth relates. Personal and direct observation, as I have mentioned, is a universal way of knowing. It combines conscious searches for patterns and explanations with unconscious mixing and synthesizing of data with memory and instinct, in combinations that vary with the individual. The scientific method is a rigorous extension of this way of knowing. By standardizing ways of observing nature and by rationalizing methods of interpretation, science tries to make reality a repeatedly demonstrable phenomenon in a framework of mechanistic cause and effect.

Like it or not we have to recognize that science itself is shot through with cultural values that aren't universally accepted. Western science assumes that mind is separate from matter and so is able to study nature dispassionately. It views nature as aspiritual, a complex linkage of causes and effects that can, if people wish, be manipulated and freely changed. We can change nature but not harm it, in this view, because no morality can be invoked in the we/it relation between person and surroundings. On a pragmatic level the science most visible to people of the North (and much of the Third World) is handmaiden to government bureaus and big resource corporations. It shouldn't be surprising that northerners— I'm thinking primarily of Natives—see science as part of a rather frightening juggernaut.[58]

A number of southern scientists and agency people have been concerned about this and have tried to bring local knowledge into the stream of decision-oriented information. From what I've seen, however, it isn't indigenous knowledge they want because they couldn't use it. Such knowledge is only meaningful within northern cultures and only useful insofar as we truly intend to make a permanent place for those cultures in our body politic. Rather, the teams of western scientists and bureaucrats want Native people to be good field assistants, reporting those sorts of observations a western naturalist or countryman would report. The knowledge is valued in proportion to its conformity to rules of western science, and we have used every northern school and university in an explicit attempt to obliterate indigenous knowing with our model and methods.

As Richard Nelson so vividly and wonderfully portrays in "Make Prayers to the Raven,"[59] Native knowledge in its original spiritual framework was a marvelously effective instrument of survival and adaptation. Why must it be given up? On the other hand, western science has produced marvelous good, if some ill as well, when used to true human purpose. Surely we can't intend to sacrifice these gains! Is there any sign of a joining of those opposites? A logical place to look might be in the new organizations springing up throughout northern Canada and Alaska which bring Native people into partnership with officialdom to resolve particular resource questions. I have not studied these closely. In large measure, these structures exist because Native northerners want something, not because they know something. In other words, through them the governments of the dominant society recognize a political interest group. The question here is, do those governments respect and seek local knowledge?

On the surface the evidence isn't encouraging. The Arctic Eskimo Whaling Commission, for example, was established to lobby for subsistence whaling for bowheads. It is now relied upon by the nation's whale management agency, the National Marine Fisheries Service, to distribute whale quotas to villages, to monitor the hunt and to report harvests. In the initial arguments over whether the traditional hunt should be shut down because of low whale numbers, Inuit estimates of bowhead populations were consistently several times those of scientists. The Inuit, it turned out, were much closer. Was this a self-serving guess that luckily proved true? Or did generations of survival-driven observation give local people an advantage in interpreting what they saw? Whatever the answer to that, it seems that in the debate since then over whale migration routes and times, number of young present, critical feeding times and places, and the prospective impacts of petroleum extraction, both the Inuit and agency people have looked to nonlocal science for definitive answers. I doubt that Inuit people have hired non-Native scientists and funded yearly scientific conferences because they believe that science is a better way of knowing. (Few Native Alaskans have chosen science as a career, and almost none who have done so are back in the northern coastal villages.) More likely, Natives sense that their knowledge is heavily discounted. Playing the game according to the rules of a more powerful polity, they fight fire with fire, data with data.

It may be, of course, that under this surface show a more subtle but powerful process of melding is going on. National Marine Fisheries Service staff stationed for months in Barrow each year may be absorbing local knowledge and coming to respect and act on it. Perhaps some sort of mixed knowing is emerging, neither purely scientific nor purely traditional. It would be good to think so.

I think most people want humanity to survive; so do I. Part of the reason, I think, is that we have a strong sense of purpose not yet achieved (or even well defined). That is what development is about. Acknowledging this urge toward ongoingness and unfolding, we should strive to promote actions that are sustainable, knowing full well that we can only choose the best of alternatives the moment provides and that we can only experiment, not conclude. We should use pulses of wealth, often from mineral and petroleum exploitation, to ease the inevitable times of scarcity through means ranging from public savings to creative vocational training to firm protection of foraging economies. I think, too, that we should experiment with stronger regional levels of social decision-making and land stewardship. And, finally, we should nurture the evolution of a new kind of knowing, a blending of science and northern tradition, because only in that language can we create a truly new kind of relation between ourselves and the North.

Part Three

Gifts

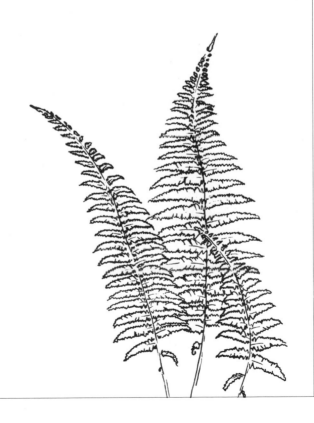

Development is a striving to reveal ever more of the potential of the person and of human communities. It is an equalization, across humanity, of opportunities for personal growth and community self-determination and well-being. It is a search for ways of living that are sustainable: that are ecologically sound, economically fruitful and socially desirable.

The primary means of development in its environmental dimension are the care of Earth, the fair and effective use and distribution of resources, and the encouragement of diversity and resilience in humanity and the rest of nature, those being key attributes of a rich, adaptive and durable humankind.

Development requires knowing the home country. It demands managerial institutions that are responsive to that knowledge. It requires a deep-felt sense of the flowing river of human endeavor: of past generations responsible for our present, and of future ones who will blame or praise us for theirs.

Hidden in these few notions are a dozen Augean stables in which generations may labor. Each will only partly achieve its particular hopes, and each new generation will redefine its hopes and, therefore, its tasks. The twin worlds of nature and culture will change, continually reorganizing the stage in which the play of human development must take place.

I spoke earlier of the need for knowledge to help the development process and of the kinds of knowing that might be brought to bear. Surely knowledge is crucial, but I am equally sure it isn't enough. Beyond the need to know more, to be more clever in our technique and technology, to be better able to manipulate human institutions and natural systems, lies something closer to purpose. In part, that something is a code of ethics, an agreed right relation of self with other, of the part to the whole.

7
An Ethic for Development

No long excursion of mine into the subject of ethics would be very useful. My answers are far too sparse, my questions too untutored. I want only to note that codes for right relations among people have a long history. They have undergone changes throughout *Homo sapiens'* span and continue to be tested and modified—seemingly at a furious pace—in our modern world of extravagant tinker technology. Similarly, and perhaps inseparably, the human sense of right relation with nature is also both ancient and rapidly evolving. This evolution is not necessarily linear or even progressive. It may be that the nature ethics of hunter-gatherers and pastoralists of distant ages were even more complex and all-embracing than our present industrial culture has yet refashioned. But with gathering momentum in the past half-century, thinkers have begun to form structures of ideas that bridge the gulf once preventing western civilization from dealing morally with nature.

Aldo Leopold's dictum for land ethics, "A thing is right when it tends to preserve the integrity, stability, and beauty of the biotic community. It is wrong when it tends otherwise"[60] wonderfully expresses my belief also, with the understanding that, like Leopold, I include humanity in that community. Land, I think, is an extension of our physical and spiritual selves. It nurtures our individual bodies and permits our social economy. It possesses our past and holds our future. It expresses the same physical reality and shares the same myths and images. It is an emanation of the same truths.

We know something is deeply amiss when farmers must plunder the soil to meet bank payments, when miners cannot collect gold without choking creeks in mud, when the rain must turn bitter poison if the lights of town are to stay lit. We know then that we have not been equal to the challenge of survival, let alone development. Soon or late we will have to pull back from the brink, rearrange our ways of doing business and hew to the task of sustaining the nurturing environmental context of human progress.

Arising both from strong self-interest and from the central fact that Earth's creatures coexist and copossess the land, a land ethic contains equal parts of acknowledgment and action. We acknowledge a destiny shared with other life, an essential commonality with other beings, and, because of what we are, a special responsibility. We act through study and stewardship, through teaching and through the creation of life-celebrating art, drama and literature.

A life of active participation with nature carries the possibility of meaningfulness, strong in its connections with place and history, replete with the rewards of service to one's self and simultaneously to others. The responsibilities of such a life are equally tangible. The worker in nature must learn Earth's messages and leave the land at least as capable of supporting and enriching life as it was at the start of the trust.

Throughout this book I have let the specific trust responsibilities of farmers, foresters, engineers, fishers, scientists, miners, teachers and resource technicians and managers remain largely implicit. For each kind of work with the land, the general responsibilities are quite obvious and nearly universal. Specific tasks of stewardship flow from the local character of the land and the particular kind of mining, farming, teaching, etc. The stewards themselves are best able to convert that knowledge into specific prescriptions for action.

The stubborn fact of our times is that only a very small minority of people have direct working contact with land, freshwaters, oceans, wildlife or natural vegetation. (Even in Alaska, the wildest, most resource-oriented American state, most people are city dwellers.) It wouldn't be enough that those few attain and live a land ethic. The urban majority, surrounded by the transformed nature we call the built environment, must do the same. Without the city person's ecological ethic, there will be no support for the necessary costs of good Earth care, either in taxes or prices paid in the market. Without it, urban folks both wrongly and dangerously will continue to feel uninvolved in this most essential sphere of human life. What needs to be done is to reconnect city residents with life by defining their participation in a land ethic.

A deep sense of personal connectedness with the land must come from experiences beyond intellectual learning, even though the role of the mind is critical. Beyond the book, the film, the lecture, the printout, the childhood stories must come hands calloused in the work of stewardship, the recollection of the smell of autumn woods, the brush of feathers across skin, the joy of a specific gift of self to the land. The kitchen garden, the annual berrying expedition, and the Peace Corps and Youth Conservation Corps tours are, for a few, familiar steps toward connectedness. We need more.

One possibility is an Earthwide requirement that all young people serve in Earth care. The idea isn't new. President Roosevelt's well-remembered Civilian Conservation Corps, though created from mixed motives, provided a choice of environmental service for young people of the 1930s. A quarter-century earlier William James wanted the nation to "conscript" youths into "an army enlisted against Nature."[61] Perhaps the time is ripe for a reemergence of this kind of program, considering both the global need and the prospect of shunting tens of billions of dollars from defense to civil purposes.

During these closing years of the 20th Century, we might be able—and surely would try—to fashion an Earth care program far less combative than James proposed and less addicted to geographic engineering than the politics of the Depression fostered. Aside from the need to show immediate and concrete results that seems inevitably to ride piggyback on all such programs, it seems to me that an even deeper-rooted tendency needs a tight rein. That is our ingrained belief that we can and should improve nature. To give one small Alaskan example, when artist-writer Rockwell Kent spent his months of quiet adventure on an island near Seward along the dark emerald coast of the central Gulf, he wrote, "The Lord must have been pleased with us today for the grand clearing up we gave this place of His... It begins to look parklike with trees stripped of limbs ten or twelve feet from the ground and the mossy floor beneath swept clean."[62] Rene Dubos expands that theme to thousands of years and whole civilizations, arguing that we can change natural systems for their own good as well as our own.[63]

Whether we can improve nature probably comes down to a question of who gives out the grades. There is no doubt, however, that by now we have done so much damage to nature (that is, exterminated or decimated species, broadcast toxins into the atmosphere and the envelope of life and soil, lowered soil fertility and reduced biological productivity) that the work of restoration—reinhabitation, to use Peter Berg and Ray Dasmann's word[64]—could occupy us for a long while.

Even the best-motivated, best-run national hands-on program still is less than enough. You cannot take time out from life to practice a land ethic any more than you can take time out to practice religious commandments. Daily life is the true practice of an ethic. Consider, then, the common acts of city life: consuming, producing goods and wastes, investing, buying and selling, voting for candidates and public expenditures, serving as volunteers, and so on. These are verbs of environmental relation. Every facet of city life (as well as country life) has consequence to nature. Every act of city living can be held up to conscious thought and choice according to the criteria of a land ethic.

Take investing, for example. In our society this is one of three primary forces (the others being the private and government purchases) in the movement of capital into and out of specific enterprises. Just as it is no longer peculiar to think of making investments according to presumed social consequences such as punishing racial discrimination, it should become commonplace to sift potential investments according to stewardship's criteria. This requires that different kinds of information than we now are accustomed to be made widely available about companies, but when was it ever easier and cheaper for humanity to have that information? Street side computer terminals could provide everything needed for firm-to-firm comparisons. Already, there are dozens of investment brokers and mutual fund managers who specialize in profitable, environmentally sound, double-green portfolios. Following in the wake of the Exxon oil spill, environmental leaders in 1989 developed the Valdez Principles, guidelines for judging environmentally sound corporations. From the standpoint of the stock-issuing company, there are beginning to be clear and measurable advantages to a good environmental record in a world where investors know, care and act.

The same is true for consuming: stewardship-conscious consumer's guides already are making ethical consumption widely possible. The day could even come when understanding clear but distant ecological connections is a habit of the public mind. We've already begun that journey. Vermont, for example, in late 1989 (and New York in 1991) debated whether to contract for electricity from Quebec until that province studied the environmental and social costs of damming more James Bay rivers. In Europe the mass marketing of environmentally benign products—so called by their makers—is common, and major retail firms in the U.S. have picked up the idea. Chances of fraud aside, it is good to know that advertisers think public interest in such products is significant.

One of the many effects of a society-wide land ethic would be a new respect for the world of labor. In living nature all work is equally critical because it contributes to the passage of energy and nutrients among species on which all depend: to that unhurried spiral from nameable form to anonymous disintegration to the gathering up again into new forms of being. Survival is the reward for good work not only for the individual for its appointed span and for the species in its longer sway, but for the larger communal system. Given that humans possess the capacity to respect, it is inexplicable that the myriad tasks that assure our survival are not respected. We scoff at dirt farmers, are cynical of lawyers, have insulting names for keepers of community security, derogate collectors of waste. Teachers are eggheads, builders are rednecks, successful business people must have sinned and unsuccessful ones must be stupid. If you and I rob each other of esteem for our roles in the human community, how can we respect a plant or animal for its work in the biotic one? Is it an accident, I wonder, that among societies such as the traditional Athabascan, where every-one of every age has a respected role, people extend respect to every creature with whom they shared the sun?

Creating Places

> A severed hand
> Is an ugly thing, and man dissevered from the earth and
> stars and his history...for contemplation or in fact...
> Often appears atrociously ugly. Integrity is wholeness,
> the greatest beauty is
> Organic wholeness, the wholeness of life and things,
> the divine beauty of the universe. Love that, not man
> Apart from that, or else you will share man's pitiful
> confusions, or drown in despair when his days darken.[65]

In those words Robinson Jeffers placed people in nature and warned of the penalties of severance from "the wholeness of life and things." Many have written about "place," which is at once created by this wholeness and is an expression of it. It is this sense of place that I offer as a transcendent phenomenon arising from the process of development and resulting in the adoption of a land ethic—or, rather, a life ethic.

Place probably has to be defined many times, each time illumi-nating one dimension while shadowing others. If one insists on brevity, there may be no better attempt than Yi-Fu Tuan's comment that "Place is a center of meaning constructed by experience."[66] The center is a geographic location, for place in this context must have a material manifestation. It is, however, a location humanized by the presence of at least one person, and by the interactivity of environs, perception and experience. The meaning of a place, the defining element distinguishing place from locus or point-in-space, comes into being when the person comprehends that there has been an exchange between his or her life and character and that of the land or place.

(I have a special place of boyhood. Perhaps nine, I broke away from the walls of New England stone keeping me civil, and found across decaying pastures and another forgotten wall a hollow enlivened by water. There were Jack-Pulpits there and green skunk-cabbage turgid with swamp water. The muck smelled like skeletons dug from Finnish fens. I recall the light playing on ash leaves, the humming silence, the atavistic feel of a solitude prohibited by towns and feared by farms. I went back again and again, building small ephemeral dams, teasing frogs or simply listening to the genius of the place, something nowhere seen but everywhere heard breathing.)

If place is a uniquely individual construct, there must be at least five billion potential places on Earth, a minimum of one per human. But one-of-a-kind places begin to merge with collective places and to acquire depth and social meaning. Just as paired aerial pictures of a tree look slightly different when seen by first one eye and then the other, but take on a third dimension and a new liveliness when merged under a stereoscope, so places can be shared by more than one experiencer and emerge to a human community as a new whole. This happens in several ways. A landscape, especially a strong one, over time captures within its circumference those people who instinctively respond to it. Then, too, landscapes involved in repeated and progressive human experiences themselves begin to change in physically measurable as well as in more subtle ways. Finally, in literature and art and in every form and record of human communication, the character of a place comes to have a historic, cultural and literary being as well as a physical one. Ultimately, the two become inseparable.

I often wonder, and can't decide, whether Alaska is a place in this sense. I tend to think it is not one center of meaning constructed out of experiences of a single community, but several. The place "Alaska" is more device than meaning, a figment of visiting writers', gold rushers' and tour advertisers' imaginations. There is a good deal more honesty in the several regional Alaskas, because it is at that smaller scale that people have intensely experienced the deep sense of place. There are regional Alaskas of the indigenous cultures, places I can never know and that are changing into something else even for their creators. And there are the places that have known and been known by non-Native immigrants for several generations. This process of land-becoming-place is nowhere complete or whole or at a resting point. Overwhelmingly, we speak *about* Alaska, not out of its bosom.

One universally necessary element in the creation of a collective sense of place is time. Native Alaska and its created places were the result of ample time. Stegner wrote in "The Sound of Mountain Water" that even the West, a good deal older in Euro-American experiences than Alaska, suffers from having been known too short a time. "One of the deprivations of people in Western America," he commented, "is that time in their country still is not molded by human living in the forms of sanctuary, continuity, and confidence that it is the ambition of all cultures to create."[67] E. J. Mishan, in "On Making the Future Safe for Mankind," expressed similar beliefs. He wrote

> ...a society congenial to man is one that strengthens his roots in earth and makes him part again of that eternal rhythm of nature in which there is time enough for things to grow slowly; in which there is time enough for trust between people to form; in which there is time to learn to care, and time to wonder and perceive beauty.[68]

Accompanying time is commitment, partly expressed as residence. We can, I suppose, think of sites where we work as places. Sites of vacation or play likewise are places. But for a community and its environs or a landscape or region to be created as collectively perceived places, they must encompass all the elements of human living. A community that houses too large a flood of transients must, at best, take much longer to settle upon its true localized and unique character, as opposed to the portable character of the stopover places of the suitcase set. This certainly is a problem in many parts of Alaska.

However, the problem of the transient subculture has to be faced. There is perhaps too great a tendency for contemporary writers to assume that transients either prevent the genius of a place from being born, or destroy that genius already in existence when they arrive. Logically, they look to rural backwaters as the dwelling of placeness, and plead that our nation must tear up its plane tickets, lock the RV in the garage and settle into a new era of rooted living. A sufficient core of rootedness is essential if the spirit of a place is to be robust, of true character and adaptive. Still, the comings and goings of people need not be destructive. Not everyone in a community has to be a permanent resident. Writing in 1966, Joseph Sonnefeld[69] suggested that there are native (root-seeking) and nonnative (peripatetic) personality types: people who are best fulfilled as rooted residents, and people who revel in change and travel. He postulates that it is the native or home-loving folks who will usually try hardest to develop a community's special character, and it is for and by those people that such work should be done. The transient personality, who delights in novelty, will enjoy the distinctiveness of the place to which he or she has come. Its character may even captivate the wanderer permanently; or if he moves on to some new excitement, at least he has a vigorous memory to cherish.

Most of us, I imagine, are mixtures of the native and nonnative personalities, possessing as we all do the typical ambiguity of humankind. A place with a robust spirit attracts, but may smother. As with a garden, a region or landscape may provide strength, surety and security, an anchor in a world of vacuous standardization; yet, it may become too enveloping, too dominating. We may then have to break free to assert our own selves, for a short or a long time, in a new place. If we were to take on a bioregionalist's love of home country, we would have to guard against a constricting provinciality. The traveler's role could be a critical one.

I have wandered through an odd country of ideas. I can only hope their connections are real and visible. Development is the central work of society, the revelation and achievement of individual and community potentials. That work is inevitably ecological because we cannot escape our existence as part of Earth's living community. Progress rests on an adequate life ethic, or land ethic, which tells us when we are in right relation to each other and our world. As our potentials unfold within this right relation, we create a sense of place so strong, so necessary, and so broadly interactive, that we and our home place become inseparable. For this to happen requires time and the commitment of residency—and one more thing: the giving of gifts.

An Exchange of Sacred Gifts

The road to my home climbs steadily along a birch-clad north-facing slope out of the valley where Fairbanks sprawls. Then, curving, it ends at the head of a fold in the hills among the aspen, facing southeast. A longtime neighbor who owns land sloping northward recently subdivided part of his holdings, meeting borough subdivision requirements with an access lane called Bloomingwoods Drive. It is in a deep trench cut through the loess, arrowing up the hill for 100 yards before hairpinning and dividing. The earth from the trench-like road cut—over three yards deep in places—was spread under the trees to the side of the lane. That was in 1986. Now the wounds are ulcerating. The ditches beside the lane are eroding, the brown mud deepening where the lane meets the public road. The sheer cutbanks of loess are calving, a prelude to major slumping to come. Along Bloomingwoods Drive the 50-year-old birches, their roots smothered by waste dirt, have died.

A bit farther up the road, where it ends in aspen, my wife and I hired a man over a quarter-century ago to backfill our new foundation and to prepare ground for a lawn. To save money we had cleared the driveway path with axe and Swede saw. We were surprised and pleased when, instead of a mildly sarcastic comment on the bumpy, root-strewn driveway, he said "Good start. No need to use a cat on the slope. Just put mine tailings over the ground the way it is, and you'll have a solid road." Then he started moving earth. Knowing nothing of the art of bulldozing, we could only trust that no matter how dubious it seemed along the way, everything would be as we wanted it when he was done. Almost by magic, and seemingly without ever looking as he backed and filled, he missed by the span of a finger or two both our raw basement walls and the trees nearby we wanted to save. All of the freighted soil was kept in a compact area, and none was spread unnecessarily over tree roots. He sloped the land laterally away from the uphill side of the house site so that snowmelt and summer rains would slide away before reaching our foundation. The inescapable cutbank on the upslope side he graded to a shallow angle that we could hold securely with perennials. "Farther up the hill, just into the permanent woods, you might dig a little ditch, not deep, across the slope to lead any runoff away from your garage and house," he remarked. And we did.

We didn't realize it then, but we now know what a fine gift the man with his Cat gave us. He gave, to us and to the land, his knowledge of wind-blown Pleistocene loess, of south slopes and sudden spring snowmelt, of the breathing of roots. For 30 years our driveway has been solid and serviceable, with negligible maintenance. Flowers and shrubs firmly hold the upslope bank in place. Only the snow that rumbles from our cabin roof to announce spring with a definitive thump ever has a chance to trickle along the foundation of the house. The only trees lost from the dozen we asked to be saved are those that, grown full in our uncrowded grounds, later shaded us and our garden too much for our good and their own.

Our earthmover had a talent, a gift of coordination, space-sense and a feel for the strength of machines and the heft of soil. The work he did was partly paid for in a market transaction, but it was also a gift, the man's gift transmitted by the work onto our environs. We have used and enjoyed it, and by our additional effort magnified it; the enlarged gift will someday be passed on to someone else. Because the original work was a gift as well as a commodity, and because we recognized it (though imperfectly then) as such, the giving created a relationship between us. For years afterward we would meet him by chance in our small town, and we would always stop to tell him how his work had lasted and what we were doing to carry on the home-making project.

This simple story contains the basic phenomenon of gift exchange so beautifully described by Lewis Hyde in "The Gift: Imagination and the Erotic Life of Property."[70] A work grows out of a gift and is passed on. The gift is received and recognized, and in new form is passed on again, to circulate forever in the community formed by the gift relationship.

It strikes me that what separates true and lasting development from the spurious, ephemeral "projects" that pass for it is this element of the gift. The frontier, I think, is a time and place where taking far outpaces giving. We come into the country to wrest something from it, and the land is judged good or bad by the value of that which can be taken. The town of immigrants is, like the country, unsettled; people come, take and go away again. Because there is little commitment to the land or town, transactions are commodity exchanges, dollar for dollar, with no added penumbra of gift. Only among a few is this gift given and received and passed along, and those few give the town what community it has. The frontier is fairly well behind us in Alaska, but only recently. The sense of community created by mutual giving is here, uneven, struggling. It often seems overwhelmed by the urge to take. During the years of pipeline construction in Alaska, and in the years of huge state-sponsored construction projects thereafter, the sense of short-term taking was overwhelming. In the land of opportunity, the opportunity was to make it, take it, keep it, not to build a community of circulating gifts.

It is important to distinguish the giving of gifts from both the sale of a good and investing in expectation of return. The dozer operator gave us more than we paid for (he could have done a much more careless and quicker job, and we still would have paid him), and he did it without expectation of future work (which we never have needed, in part because of his care). Today it is commonplace for public and private agencies to "think of the future," by which they mean doing something now whose payoff will come later. A farmer spreads fertilizer in spring in an amount neatly calculated to pay him back in fall harvests; putting on fertilizer beyond the marginal rate of return would be nonsensical. A logging company may be required by law to plant trees to replace those cut from public land. The regulation is society's way of spanning human horizons of self-interest, necessary because trees grow so slowly that the planter cannot be the harvester. Or in a third example, a state fish and game manager spends funds on a prescribed fire, expecting that a future moose hunter will benefit. None of these is real giving—though within the bounds of utility they are wise, even far-sighted.

This sort of expanded self-interested is highly worthwhile, far better than the individual self-interest so easily found around us. It can be a grounds for saving endangered species and essentially every other act of good Earth care. With that alone as a guide, we can walk much farther along the way than we yet have come. Still, utility, prudence and the investment principle can only help us survive as a species on this planet. It assures nothing more. Utility deals with bread, and bread is necessary but not sufficient. Giving, on the other hand, deals with what we do with the bread, with survival. That is why I think giving is tied so closely to development, which, starting with survival, is the story of what comes of it.

William Everson, the Santa Cruz poet, wrote in "Birth of a Poet"[71] that a person requires three things for a fulfilled life: a sacred calling, a sacred love and a sacred country. In each case the word sacred sets the noun apart from the ordinary, even from the repeatable. The strong suggestion is that though a person may have several callings, loves or countries, only one can evoke the deepest dedication and commitment, and can match with its own the offered spirit of the person. Everson understands "sacred country" within the Christian context, but the idea is universal. For the Hopi and Navaho fighting coal mining and nuclear power plants, the sacred country has both general meaning (the Earth itself, as a whole, is sacred) and specific identity as places where the enspiritedness of the world is especially powerful.[72] To them, a place is sacred when revelations occur there, when you make offerings to it, and when all your skills and ceremonies stem from and are bent toward it. A sacred place has inseparable parts. A sacred country explains you to yourself and others, and once committed to it, your life is no longer portable: if you leave, something important stays behind.

Native northerners lived in country sacred to them. This river of belief may still run deep, much deeper than we of the busy and largely indifferent majority can see. I hope so, because the greatest lesson we still must learn is the giving of sacred gifts to the country, hence ourselves. Lacking it, our science will continue to produce answers in search of questions, doors needing rooms they can reveal. Our environmental managers will continue to lose games of brinkmanship. Our political leaders will continue, unknowingly, to offer illusion instead of vision. And we, as persons, will continue our eyeless search for meaning. I am not suggesting that we all adopt the traditional spirit system of the Native people. Even if we come eventually close to it, we must start where we are and fashion our own map into the sacred country so that we know where we are when we arrive at its borders.

With apologies to Robert Frost, I offer a version of "The Gift Outright" for Alaska:

This now, is that ultimate westward place
Where we must stop at arctic shores and search
Behind the eyes for new and vast frontiers.
Transient strangers here, and deaf to all
The messages from frozen tender earth
We want too much and give too little back.
We three whose hands must clasp stand far apart:
One knows the Old Way, signed with rock-tipped spear
A Declaration of Dependency
But fumbles now with tools that slash the heart;
Another knows the world but not the land,
Has more to learn than time to learn it well.
The third is Earth, the Mother who will care
For those who stay to make their home with her.
In dreams of wealth our nation almost died;
Perhaps in facing truth we will survive.

Glimmerings

M oving on its slightly drunken elliptical path, the Earth presents a cold shoulder to the sun most of the year. Alaska
rides there on that northern shoulder, luckily surrounded by
ameliorating seas but still colder, still farther from the sun's full
glance, than most other parts of the planet. The mantle of life is here,
an interlocking system of successful attempts—good ideas, you
could call them, or species if you are a biologist—to catch and use
the pale, tangential rays of the Warmth-Giver. Cold facts are
inescapable, and thus living systems in the North's uplands, streams
and polar oceans transform less energy per year than those of the
South. The tactics of success are as numerous as the species that
express them, but the main strategic patterns are few. Energy,
being scarce overall, must be conserved; being highly seasonal,
it must be banked for use in the dark and frozen months or for
quick spring starts. The times when and places where energy and
nutrients are more liberally available are rather far apart, like the
few years in a decade when the eaters of spruce seeds have a feast,
or the rare occasions when, by chance, abundant browse, light
winter snowfall and temporary lulls in predator pressures allow
moose to raise twins to maturity. And so strategies of opportunism,
high mobility, pioneering capacity and niche generalization are
rewarded by survival.

Hunting societies from Asia were already proven in northern
testing grounds when they became North America's first pioneers.
Eventually their progeny pushed through the whole western hemisphere, evolving variants better suited to regional conditions as
different as the deserts of Arizona and the shrieking coast of Tierra
del Fuego. When Europeans and Euro-Americans came to Alaska,
they were met by abundant proof of the durability of foraging
strategies.

The newcomers came to plant their industries onto still another
part of the globe. Whether industrialization, as an idea for human
living, will endure for very long is moot. The jury is still out, and
rumors are that there is no consensus. Very possibly the expanding
human population and high per-capita consumption on which the

whole experiment rests are untenable. Still, there is no reason to believe that hunting-gathering is the only viable solution to the problem of northern survival. Something valuable, surely, can be done with the notion that work can be traded for money, and money for leisure and an infinite variety of good things. Some wise use, surely, can be found for the idea of nature as clockwork, and humans as clock-repairers, even if contemporary science and technology, the present outcome of that idea, have their dubious and mortally dangerous aspects.

Whatever the new idea for northern living turns out to be, it must take care of humans as physical entities. We must live and build communities somewhere, eat something, communicate, mine and grow materials. In any future except annihilation, we will create a human habitat; that habitat will serve best, I think, that violates least the nature out of which it is created, in which it is imbedded, and from which it is nurtured. It will be northern in style (by which I mean very different from the roughly transplanted southern skills and institutions with which we now make do), reflecting the strategies that creatures green and brown have proven viable through the ages. We will conserve energy and materials and store them for predictable and unexpected hard times. We will harvest animals and plants at rates suited to slow northern replacement. We will develop a new balance of generalists and specialists in our society and its economy; we will be opportunistic and resilient. I doubt that this new way of living will depend on intricate, comprehensive, detailed planning and the gathering of the infinite bits of real-time information such a planning system would require. That kind of strategy, so appealing to the ultrarational mind, is far too costly, both in money and lifetime—and, let's admit it, too liable to collapse—to be worth the effort. Instead, we will combine a kind of lighthearted and lighthanded muddling through with a yet-unknown expansion of our collective horizon of thoughtfulness. Above all, earthcare will be the primary science and concern of society. Whatever our destructive capacity in that future society (and I wager it will be less than today's), it will be exceeded by our restorative capacity.

So we will tend to our bodies and those of companion Earthriders to the necessary end of mutual survival. And then, if we are more than hands and mouths, we will tend also to the spirit. As central as that need is, I must leave it as personal work yet undone, and as a geography to be explored by those better tutored than I. Yet, we all know the spirit is mute without the body and the body is mechanical without its spirit, from which we can expect that the reenchantment of the world will involve the interdependent traffic between corporeal and incorporeal. Future enduring northern societies will possess and express a land ethic, a life ethic, of depth and scope beyond our ken because such an ethic is an essential part of that traffic. Future northerners will have come into a country made sacred by the giving and receiving of innumerable gifts. And the country will have come into them, their very blood singing the myths of lasting northern realities.

We do not dream the absolutely impossible, which destroys hope. Neither do we dream the easily achievable, which destroys meaning. The realm of dreams is the barely possible. And so I dream into futures, freely rebuilding myself and the world. But I can act only now, starting here, with this body, taking one step. If I am lucky some of the places I encounter will seem familiar.

> *And so the events of our lives move*
> *From the uncertain and unknowable future*
> *To the confused and ambiguous present*
> *And then to the memorable, receding past.*
>
> *The future must be uncertain,*
> *Else there could be no act of creation;*
> *The present must be confused,*
> *Lest we rest complacent in ignorance;*
> *The past must be full of memories,*
> *Or all our experience will have been in vain.*

Sources and Resources

Preface and Introduction

1. Wendell Berry, *Home Economics* (San Francisco: North Point Press, 1987), 146.

2. International Union for the Conservation of Nature and Natural Resources, *World Conservation Strategy: Living Resource Conservation for Sustainable Development* (Gland, Switzerland: IUCN, 1980).

Part One   This Place Called Alaska

3. Robert Frost, "The Gift Outright," in *Come In and Other Poems* (New York: Henry Holt and Co., 1943), 185.

Reading:

   Though it has no pretensions as a text, Neil Davis' *Alaska Science Nuggets*, Geophysical Institute, University of Alaska Fairbanks, 1982 (now available from the University of Alaska Press), is a far-ranging and extremely informative compendium of 400 articles originally written by various scientists for various Alaskan newspapers. There is no other single volume of such scope on Alaskan natural history and science.

Oceans of Air, Oceans of Water

4. T. C. McLuhan, *Touch the Earth* (New York: E. P. Dutton, 1971), 25.

5. Charles Hartman and Philip Johnson, *Environmental Atlas of Alaska* (Fairbanks: Institute of Water Resources, University of Alaska Fairbanks, 1978), 61.

6. Harold Searby and Ivan Branton, "Climatic Conditions in Agricultural Areas," in *Alaska's Agricultural Potential*, ARDC *Publ. No. 1* (Fairbanks: Alaska Rural Development Council, University of Alaska Fairbanks, 1974), 29-44.

7. Hartman and Johnson, op. cit., 91.

8. Hartman and Johnson, op. cit., 21.

9. Ted Dyrness, Leslie Viereck and Keith Van Cleve, "Fire in Taiga Communities of Interior Alaska," in *Forest Ecosystems in the Alaskan Taiga: A Synthesis of Structure and Function*, ed. by Van Cleve et al. (New York: Springer-Verlag, 1986), 74-86.

10. A. T. Pruter, "Development and Present Status of Bottomfish Resources in the Bering Sea," *Journal of the Fisheries Research Board of Canada*, 30(1973):2373-2385.

11. Edgar Bailey and Glenn Davenport, "Die-off of Common Murres on the Alaska Peninsula and Unimak Island," *Condor* 74(2)(1972):215-219.

12. Alan M. Springer, D. Roseneau, D. Lloyd, C. P. McRoy and E. Murphy, "Seabird Responses to Fluctuating Prey Abundance in the Eastern Bering Sea," *Marine Ecology Progress Series* 32(1986):1-12 .

13. C. Peter McRoy, John Goering and William Shiels, "Studies of Primary Production in the Eastern Bering Sea," in *Biological Oceanography of the Northern North Pacific Ocean*, ed. by Takenouti et al. (Tokoyo: Idemitsu Shoten, 1972), 199-216.

14. Anthony DeGange and Gerald Sanger, "Marine Birds," in *The Gulf of Alaska: Physical Environment and Biological Resources*, ed. by Hood and Zimmerman (Washington, D.C.: U.S. Minerals Management Service, 1987), 479-524.

Readings:

For a very broad overview of polar geography, *Arctic Environment and Resources* by Sater, Ronhovde and Van Allen (Washington, D.C.: Arctic Institute of North America, 1971, 309 pp.), is a good reference.

The 6-volume series of *Alaska Regional Profiles*, edited by L. Selkregg (Anchorage: Arctic Information and Data Center, University of Alaska Anchorage, 1974, 1975, 1976) is a much more detailed atlas of Alaskan geography, still very valuable.

A newer treatment of arctic atmospheric dynamics and air pollution problems is *Arctic Air Pollution* edited by Stonehouse (Cambridge University Press, 1986).

Permafrost and its practical consequences are well treated in Williams' *Pipelines and Permafrost: Physical Geography and Development in the Circumpolar North* (New York: Longman Inc., 1979).

The Land

15. Joint Federal-State Land Use Planning Commission for Alaska, *Major Ecosystems of Alaska* (map) (Anchorage, AK: JFSLUPC; Denver: U.S. Geological Survey, 1973).

16. Michael Spindler and Brina Kessel, "Avian Populations and Habitat Use in Interior Alaska Taiga," *Syesis* 13(1)(1980):61-104.

17. John Bryant and F. Stuart Chapin III, "Browsing-Woody Plant Interactions During Boreal Forest Plant Succession," in *Forest Ecosystems in the Alaska Taiga*, ed. by Van Cleve et al. (New York: Springer-Verlag, 1986), 230.

18. David Klein, "Postglacial Distribution Patterns of Mammals in the Southern Coastal Regions of Alaska," *Arctic* 18(1)(1965):7-20.

19. Patrick Webber, "Tundra Primary Productivity," in *Arctic and Alpine Environments*, ed. by Ives and Barry (London: Methuen and Co., 1974).

20. George Batzli, "Populations and Energetics of Small Mammals in the Tundra Ecosystem," in *Tundra Ecosystems: A Comparative Analysis*, ed. by Bliss et al. (New York: Cambridge University Press, 1981), 377-396.

21. Keith Van Cleve and John Yarie, "Interaction of Temperature, Moisture, and Soil Chemistry in Controlling Nutrient Cycling and Ecosystem Development in the Taiga of Alaska," in *Forest Ecosystems in the Alaskan Taiga*, ed. by Van Cleve et al. (New York: Springer-Verlag, 1986).

22. Richard Hurd, "Annual Tree-Litter Production by Succes-sional Forest Stands, Juneau, Alaska," *Ecology* 52(5)(1971):881-884.

Readings:

Alaska's landforms are thoroughly but readably described in *Landscapes of Alaska: Their Geological Evolution,* edited by Howell Williams (Berkeley: University of California Press, 1958).

Interior Alaska: A Journey Through Time, edited by Jean Aigner et al., Alaska Geographical Society's 1986 book, has excellent information on the natural history of that region.

The U.S. Forest Service's 7-part series entitled *The Forest Ecosystem of Southeast Alaska,* is a review of upland and fresh water natural history in the rainforest region (Portland: Pacific Northwest Forest and Range Experiment Station General Technical Reports, 1974). Probably the most thorough roundup of tundra ecology is in the volume *Tundra Ecosystems: A Comparative Analysis* (see footnote 25 above). A newer, lighter, and considerably broader treatment including fresh water, marine and Antarctic environments is Stonehouse's *Polar Ecology* (New York: Chapman and Hall, 1989).

For scientific descriptions of boreal forests in central Alaska, no other source is as comprehensive as the *Forest Ecosystems...* book cited in ref. 21 above.

Flowing Water, Still Water

23. John Hobbie, ed., "Limnology of Tundra Ponds," *U.S. Int'l. Biol. Program Synthesis Series 13* (Stroudsberg: Dowden, Hutchinson and Ross, Inc., 1980).

24. Vera Alexander and Robert Barsdate, "Physical Limnology, Chemistry, and Plant Productivity of a Taiga Lake," *Int'l. Revue Ges. Hydrobiol.* 56(6)(1971):825-872.

25. Robert Barsdate and Vera Alexander, "Geochemistry and Primary Productivity of the Tangle Lake System, an Alaskan

Alpine Watershed," *Arctic and Alpine Research* 3(1)(1971):27-41.

26. Paul Frey, Ernst Mueller and Edward Berry, *The Chena River: A Study of a Subarctic Stream* (Fairbanks: Federal Water Quality Admin., Alaska Water Laboratory, University of Alaska Fairbanks, Project 1610, 1970.)

27. James King and Calvin Lensink, *An Evaluation of Alaskan Habitat for Migratory Birds* (Washington D.C.: USDI Bureau of Sport Fisheries and Wildlife, 1971).

28. Stephen Murphy, Brina Kessel and Len Vining, "Waterfowl Populations and Limnological Characteristics of Taiga Ponds," *J. Wildl. Manage.* 48(4)(1984):1156-1163.

29. Lo-Chai Chen, "The Biology and Taxonomy of the Burbot, Lota lota leptura, in Interior Alaska," *Biological Papers No. 11, University of Alaska Fairbanks*, 1969.

30. George Van Whye and James Peck, *A Limnological Survey of Paxson and Summit Lakes in Interior Alaska*, Alaska Dept. Fish and Game Information Leaflet 124 (Juneau: ADF&G, 1968).

31. Saree Gregory, "Population Characteristics of Dolly Varden in the Tiekel River, Alaska" (M.Sc. thesis, University of Alaska Fairbanks, 1988).

32. Robert Armstrong and James Morrow, "The Dolly Varden Charr, Salvelinus malma," in *Charrs: Salmonid Fish of the Genus Salvelinus*, ed. by Balon (The Hague, Netherlands: Dr. W. Junk, 1980), 99-140.

Messages From Earth

33. Office of Inspector General, *Report on Audit of the Anchorage Office's Administration of Development Activities in the Alaska Mutual Help Home Ownership Program*, Rpt. No. 89-TS-101-0007, August 29, 1989.

34. John Kelsall and David Klein, "The State of Knowledge of the Porcupine Caribou Herd," in *Trans. North Amer. Wildlife and*

Natural Resources Conference (Washington D.C.: Wildlife Management Inst., 1979).

35. Glenn Juday, "The Rosie Creek Fire," *Agroborealis,* University of Alaska School of Agriculture and Land Resources Management, Fairbanks, 17(1)(1985):11-20.

36. Howard Feder and Stephen Jewett, "Feeding Interactions in the Eastern Bering Sea with Emphasis on the Benthos," in *The Eastern Bering Sea Shelf: Oceanography and Resources,* ed. by Hood and Calder (Washington, D.C.: National Oceanographic and Atmospheric Administration, 1981), 1229-1261.

Part Two   Learning

Gold, Grain and Slow-Grown Wood

37. Joel Garreau, *The Nine Nations of North America* (Avon Books, 1981).

38. Loren Eisely, *The Last Days,* in *Notes of An Alchemist* (New York: Charles Scribner's Sons, 1972), 80-85.

39. James Reynolds, Rodney Simmons and Alan Burkholder, "Effects of Placer Mining Discharge on Health and Food of Arctic Grayling," *Water Resources Bulletin, Paper No. 88090,* June 1989:625-635.

40. Katherine Holmes, "Natural Revegetation of Gold Dredge Tailings at Fox, Alaska" (M.Sc. thesis, University of Alaska Fairbanks, 1982).

41. James Durst, "Small Mammals in Relation to Natural Revegetation of Gold Dredge Tailings at Nyac, Alaska" (M.Sc. thesis, University of Alaska Fairbanks, 1984).

42. John Schoen, Matt Kirchhoff and Jay Hughes, "Wildlife and Old-Growth Forests in Southeast Alaska," *Natural Areas Jour.* 8(3)(1988):138-145.

43. Anonymous, "Tongass Debate Rests," in Spec. Suppl. to *Sourdough Notes,* Employee Newsletter, Alaska Region U.S. Forest Service, December 1990.

44. Wayne Thomas and Carol Lewis, "Alaska's Delta Agricultural Project: A Review and Analysis," *Agricultural Admin.* 8(1981):357-374.

Readings:

A general summary of post-statehood agriculture in Alaska, with guesses about the future, is in "Agriculture in Alaska: 1976-2000 AD" by Wayne Thomas, *Alaska Review of Business and Economic Conditions* XIII(2) (Anchorage: Institute of Social, Economic and Government Research, University of Alaska Anchorage, 1976).

Natural History magazine for August 1988 contains a series of articles on Tongass National Forest management, emphasizing ecological changes after ancient forests are cut.

Fitting Into the Country

45. Bill McKibben, *The End of Nature* (New York: Random House, 1989).

46. Walter Firey, *Man, Mind, and Land: A Theory of Resource Use* (New York: Free Press, 1960).

47. Alfred North Whitehead, *Science and the Modern World* (1925; reprint, New York: Mentor Books, 1959), 184.

48. Gordon Harrison, "A Citizen's Guide to the Constitution of the State of Alaska," *Institute of Social and Economic Research Report Series,* No. 55, (1982):67-76.

49. Raymond Dasmann, John Milton and Peter Freeman, *Ecological Principles for Economic Development* (Morges, Switzerland: International Union for Conservation of Nature and Natural Resources; Washington, D.C.: Conservation Foundation, John Wiley and Sons, 1973).

Toward Enduring Societies

50. Lester Brown, *Building a Sustainable Society* (New York: W.W. Norton, 1980).

51. World Commission on Environment and Development, *Our Common Future* (New York: Oxford University Press, 1987).

52. Michael Robinson, Michael Pretes and Wanda Wuttunee, "Investment Strategies for Northern Cash Windfalls: Learning From the Alaska Experience," *Arctic* 42(3)(1989):265-276.

and

Michael Pretes and Michael Robinson, "Beyond Boom and Bust: A Strategy for Creating Sustainable Development in the North," *Polar Record* 25(153)(1989):115-120.

53. David Ross and Peter Usher, *From the Roots Up: Economic Development as if Community Mattered* (New York: The Bootstrap Press, Croton-on-Hudson, 1986).

54. Alaska Department of Fish and Game, "Alaska's Per Capita Harvests of Wild Food," *Alaska Fish and Game* 21(6)(1989):14-15.

55. Charles H. W. Foster, *Experiments in Bioregionalism: The New England River Basins Story* (Hanover: University Press of New England, 1984).

56. Kirkpatrick Sale, *Dwellers in the Land: The Bioregional Vision* (San Francisco: Sierra Club Books, 1985).

57. Barry Lopez, *Arctic Dreams: Imagination and Desire in a Northern Landscape* (New York: Charles Scribner's Sons, 1986). (Especially "The Country of the Mind," pp. 252-301.)

58. Donald Gamble, "Crushing of Cultures: Western Applied Science in Northern Societies," *Arctic* 39(1)(1986):20-23.

59. Richard Nelson, *Make Prayers to the Raven* (Chicago: University of Chicago Press, 1983).

Part Three Gifts

An Ethic For Development

60. Aldo Leopold, *A Sand County Almanac and Sketches Here and There* (New York: Oxford University Press, 1949).

61. William James, *The Moral Equivalent of War*, in *Memories and Studies* (New York: Longmans, Green Co., 1911).

62. Rockwell Kent, *Wilderness: A Journal of Quiet Adventure in Alaska* (New York: Blue Ribbon Books, 1920), 35.

63. Rene Dubos, *The Wooing of Earth: New Perspective on Man's Use of Nature* (New York: Charles Scribner's Sons, 1980), 79.

64. Peter Berg and Raymond Dasmann, *Reinhabiting a Separate Country*, ed. by Peter Berg (San Francisco: Planet Drum Foundation, 1978).

65. Robinson Jeffers, "The Answer," in *The Selected Poetry of Robinson Jeffers* (New York: Random House, 1959), 594.

66. Yi-Fu Tuan, "Place: An Experiential Perspective," *The Geographical Review*, LXV(2)(1975):151-165.

67. Wallace Stegner, *The Sound of Mountain Water* (New York: Doubleday, 1969).

68. E. J. Mishan, "On Making the Future Safe for Mankind," *Public Interest No. 24*, Summer 1971.

69. Joseph Sonnefeld, "Variable Values in Space and Landscape: An Inquiry into the Nature of Environmental Necessity," *Journal of Social Issues* XXII (4)(1966):71-82.

70. Lewis Hyde, *The Gift: Imagination and the Erotic Life of Property* (New York: Vintage Books, 1979).

71. William Everson, *Birth of a Poet: The Santa Cruz Meditations* (Santa Cruz: Black Sparrow Press, 1982).

72. Trebbe Johnson, "Between Sacred Mountains," *Amicus Journal,* Summer (1987):18-23.

Readings:

The literature on environmental ethics is big and growing. The University of Georgia has published a quarterly journal called "Environmental Ethics" since 1979. In 1989 Donald Davis did us a service by writing an annotated bibliography of writings on the subject, *Ecophilosophy: A Field Guide to the Literature* (San Pedro: R. and E. Miles, 1989).

For an excellent interpretation of Aldo Leopold's work, see *Companion to A Sand County Almanac: Interpretive and Critical Essays,* ed. by J. Baird Callicott (Madison: University of Wisconsin Press, 1987).

Economics as if the Earth Really Mattered: A Catalyst Guide to Socially Conscious Investing, by Susan Meeker-Lowry (Philadelphia: New Society Publishers, 1988), is one example of books available on combining ethics and investment practice.

Index